**CROSS SECTION OF
THE UNITED STATES OF AMERICA**

# U.S. 40

### CROSS SECTION OF

# THE UNITED STATES OF AMERICA

## by George R. Stewart
### maps by Erwin Raisz

GREENWOOD PRESS, PUBLISHERS
WESTPORT, CONNECTICUT

**Library of Congress Cataloging in Publication Data**

Stewart, George Rippey, 1895-
   U.S. 40: cross section of the United States of America.

   Reprint of the 1953 ed.
   1. United States Highway 40.   2. United States—Description and travel—1940-1960.   I. Title.
E169.02.S843   1973        917.3'04'92        72-11338
ISBN 0-8371-6655-1

Originally published in 1953 by the Houghton Mifflin Company, Boston

Copyright 1953 by George R. Stewart

All rights reserved

Reprinted with the permission of the Houghton Mifflin Company

Reprinted in 1973 by Greenwood Press
A division of Congressional Information Service
88 Post Road West, Westport, Connecticut 06881

Library of Congress Catalog Card Number 72-11338

ISBN 0-8371-6655-1

Printed in the United States of America

10 9 8 7 6 5 4 3 2

# CONTENTS

### The Background

| | |
|---|---|
| I. Two-way Cross Section | 1 |
| II. How U. S. 40 Began | 10 |
| III. Roads in America | 17 |
| IV. On Motoring | 24 |

### The Pictures

| | |
|---|---|
| I. As for This Book— | 33 |
| II. Jersey Prologue: Atlantic City to New Castle | 36 |
|     1. BEGINNING OF U. S. 40 | 40 |
|     2. COASTAL PLAIN | 42 |
|     3. FARMHOUSE | 44 |
| III. Post Road: New Castle to Baltimore | 46 |
|     4. NEW CASTLE | 58 |
|     5. SIX-LANE HIGHWAY | 60 |
|     6. BUSH RIVER | 62 |
|     7. BALTIMORE ROWS | 64 |
| IV. Bank Road and Turnpike: Baltimore to Cumberland | 66 |
|     8. ELLICOTT CITY | 72 |
|     9. FREDERICK | 74 |
|     10. MARYLAND COUNTRYSIDE | 76 |
|     11. HORRIBLE EXAMPLE | 78 |
|     12. MOUNT PROSPECT | 80 |
|     13. RIDGE AND VALLEY | 82 |

## V. National Road: Cumberland to Wheeling     84

    14. THE NARROWS     96
    15. FROM LITTLE SAVAGE MOUNTAIN     98
    16. MASON-DIXON LINE     100
    17. FORT NECESSITY     102
    18. BRADDOCK'S GRAVE     104
    19. TOLL HOUSE     106
    20. COAL MINE     108
    21. WHEELING     110

## VI. National Road, Extended: Wheeling to St. Louis     112

    22. CAMBRIDGE, OHIO     122
    23. S-BRIDGE     124
    24. HIGHWAY AND TREE     126
    25. MILEPOSTS     128
    26. TAVERN     130
    27. TAYLORSVILLE DAM     132
    28. IN FULL GLORY     134
    29. FARM ON THE NATIONAL ROAD     136
    30. VICTORIAN ELEGANCE     138
    31. ROADSIDE VENDOR     140
    32. BENJAMIN HARRISON ERA     142
    33. VANDALIA     144

## VII. Boonslick, and Modern: St. Louis to Kansas City     146

    34. MISSISSIPPI RIVER     150
    35. ROAD-MAKING     152
    36. BOONVILLE     154
    37. DOUBLE HIGHWAY     156
    38. KANSAS CITY     158
    39. SIGN POST     160

## VIII. Smoky Hill Trail: Kansas City to Denver     162

    40. HAYFIELD     170
    41. BIT OF THE OLD WEST     172
    42. KANSAS CORN     174
    43. ROADSIDE PARK     176
    44. PLAINS BORDER     178
    45. SHADED STREET     180
    46. GRAINFIELD     182

## VIII. Smoky Hill Trail: Kansas City to Denver *(continued)*

| | | |
|---|---|---|
| 47. | HIGH PLAINS | 184 |
| 48. | ROUTES DIVIDE | 186 |

## IX. Berthoud's Road: Denver to Salt Lake City — 188

| | | |
|---|---|---|
| 49. | FRONT RANGE AND HOGBACK | 194 |
| 50. | TWO SPECIES | 196 |
| 51. | IDAHO SPRINGS | 198 |
| 52. | BERTHOUD PASS—EASTERN APPROACH | 200 |
| 53. | BERTHOUD PASS—EASTERN ASCENT | 202 |
| 54. | CONTINENTAL DIVIDE | 204 |
| 55. | BERTHOUD PASS—WESTERN APPROACH | 206 |
| 56. | BYERS CANYON | 208 |
| 57. | MEADOW AMONG MOUNTAINS | 210 |
| 58. | THE RABBIT EARS | 212 |
| 59. | STREAM IN SNOW | 214 |
| 60. | OATFIELD | 216 |
| 61. | SHEEPHERDER | 218 |
| 62. | ROCK WALL | 220 |
| 63. | STATE BORDER | 222 |
| 64. | ROOSEVELT, UTAH | 224 |
| 65. | PLATEAU COUNTRY | 226 |
| 66. | TARGET OF OPPORTUNITY | 228 |
| 67. | BEAVER DAMS | 230 |
| 68. | VALLEY IN THE WASATCH | 232 |

## X. Cutoff and California Trail: Salt Lake City to Reno — 234

| | | |
|---|---|---|
| 69. | GREAT SALT LAKE AND SMELTER | 242 |
| 70. | SALT-FLAT | 244 |
| 71. | WENDOVER | 246 |
| 72. | THE GREAT BASIN | 248 |
| 73. | PEQUOP SUMMIT | 250 |
| 74. | SUN ON HIGHWAY | 252 |
| 75. | EMIGRANT PASS | 254 |
| 76. | TIE-HOUSE | 256 |
| 77. | PLAYA | 258 |
| 78. | GOLCONDA SUMMIT | 260 |
| 79. | IMLAY | 262 |
| 80. | THE FORTY-MILE DESERT | 264 |
| 81. | LAHONTAN STORY | 266 |

XI. Truckee Route: Reno to San Francisco    268

    82. FIRST PINES    276
    83. TRUCKEE CANYON    278
    84. HIGHWAY AND RIVER    280
    85. DONNER PASS    282
    86. SNOW SCENE    284
    87. FOREST PRIMEVAL    286
    88. EMIGRANT GAP    288
    89. THE DIGGINGS    290
    90. COAST RANGE    292
    91. SAN FRANCISCO BAY BRIDGE    294
    92. END 40    296

XII. A Few Reflections    298

    The Names    303
    Author's Note    310

**THE BACKGROUND**

# TWO-WAY CROSS SECTION

"Ladies and gentlemen!"—to borrow a formula from the conductor of the sightseeing tour—"You are now about to pass along U. S. 40. You may observe, on your right hand, and on your left—the United States of America!"

U. S. 40 runs from the Atlantic Coast of New Jersey to the Pacific Coast of California. Both in space and in time it provides a cross section of the nation. . . .

This is a book about that road—in part a history of how it has come to be what it now is, but more a presentation, with the aid of photographs, of its existing reality.

"What is U. S. 40?" With anything so well known and obvious, no one would consider, at first thought, that this question could raise difficulty. Anyone might reply: "Why, it is merely a road you take—east or west—across the country! It goes through Baltimore, Columbus, St. Louis, Kansas City, Denver, Salt Lake, and San Francisco." But, as with so many simple and well-known things, the question of definition, more closely considered, becomes difficult.

In the first place, U. S. 40 has no status by law. According to an official bulletin of the Public Roads Administration, "the designation as a U S route is without legal or administrative significance," and such routes are designated and numbered merely "for the convenience of map makers, information services, and highway travelers." No one even seems to be sure whether the correct usage is "U. S. Highway 40" or "U. S. Route 40." Even the official practice differs, and some state highway departments merely fall back upon the colloquial "U. S. 40."

There will obviously be, moreover, some difference in our attitude, depending upon whether we think in terms of "highway" or of "route."

Highway suggests the physical—the ocean-to-ocean slab of pavement, now expanding into four or six lanes, now contracting to two. Mostly it shows either the gray-white of concrete or the black of some bituminous surface; occasionally, the rich pink-red of brick. It includes hundreds of bridges and culverts, scores of underpasses and overpasses, two tunnels, and thousands of signposts. Even its area is considerable, for if its right-of-way were lifted up, straightened, clipped into ten-mile lengths, and laid down again with the lengths side by side, a rectangle about ten miles long and seven miles wide would be formed, an area three times as great as Manhattan Island. To use the crassest of measuring devices, the dollar-value of U. S. 40, even without counting the value of its right-of-way, must amount to several hundred million dollars.

Even considered merely as this physical thing, U. S. 40 is far from simple. It is at this writing 3091 miles long from end to end. But this figure will be different next year, because of relocations. There are, furthermore, some two hundred additional miles of the road labeled Alternate, By-pass, Business Route, or Temporary Route. On the other hand there are several hundred miles of pavement to which U. S. 40 does not possess exclusive right, but which it shares with some state highway or some other national highway.

But if we consider U. S. 40 as a route, not as a highway, its actuality is even more subtle. For, with a route, we deal with something that cannot be expressed in terms of concrete and asphalt and brick or valued in dollars, but is of the nature of an abstraction, like a line in geometry, connecting one point with another, expressible only in terms of distance and direction.

To relocate a highway, one must call on men and machines for physical work, but to relocate a route one need only draw a line upon a map, or make a decision within a mind.

As highway, U. S. 40 may be complex as the body of a man is complex —developing or deteriorating, always changing. But as route it is complex after the manner of a man's personality, in that it cannot be weighed, or even defined clearly—though it is highly interesting, and can be talked about indefinitely.

Yet, as far as this book is concerned, we need not belabor this distinction. Here we consider U. S. 40 *both* as a highway *and* as a route.

In the photographs you see necessarily the highway, the physical thing in space, translatable into light-rays! It loops ribbon-like among the hills;

it pauses at a valley-rim and drops into a ten-mile straightaway; it switches back and forth up the face of a pass; it springs out into space across a high bridge; pestered by stop lights, like a stream impeded by rocks and dams, it forces a way along some city street.

But, writing, we deal more with a route. We consider a quite unphotographable and almost ineffable continuity, in time as well as in space—a something that may first have been a buffalo-trail or an Indian-path, and then a pack-horse way and a cart-road, and may at last have become a four-lane freeway—although at no time, in spite of innumerable relocations, did it ever cease to be the route by which one passed from the same point to the other same point.

Nay more—in this book, U.S. 40 must be even more than a highway *and* a route. For we must consider all that it means to the man who drives along it. It must be not only what can be seen, but also what can be felt and heard and smelled. We must concern ourselves with the land that lies beside it and the clouds that float above it and the streams that flow beneath its bridges. We must remember the people who pass along it, and those others who passed that way in the former years. We can forget neither the ancient trees that shadow it, nor the roadside weeds that grow upon its shoulders. We must not reject the wires that parallel it, or the billboards that flaunt themselves along its margins. We must accept the slums of "Truck Route" as well as the skyscrapers of "City Route," and the fine churches and houses of "Alternate Route." We must not avert our eyes even from the effluvia of the highway itself—the broken tires, and rusting beer cans, and smashed jack-rabbits. Only by considering it all, as we drive from the east or from the west, shall we come to know in cross section, the United States of America....

First and most obviously, U.S. 40 cuts a cross section in space. It traverses the breadth of twelve states and corners of two others. It passes through 105 counties. Of our thirty most populous cities, eight lie upon its route. Of smaller towns and crossroads-settlements, the number rises into the hundreds.

If you would rather turn your attention to the countryside, you find yourself passing through the General Farming Belt and the Little Corn Belt and the Wheat Belt and the Cattle Belt. In the fields that border U.S. 40 you see every important field crop grown in the United States, cotton only excepted. You also pass coal mines, and traverse oil fields, and see the smoke hanging heavy about the high chimneys of copper-smelters. As for manufacturing plants, you see premises from which come out practically every product that our civilization uses.

Your interest, however, may turn more to the natural background—to the continent rather than the nation. Here also U. S. 40 presents a cross section. It traverses most of the great topographical divisions—the Atlantic Plain, Appalachian Highlands, Interior Plains, Rocky Mountains, Intermontane Plateaus, and Pacific Mountains. In so doing it passes in geological review rocks of all ages from Archaean to Recent. You will also see the soils that come from the weathering of those rocks—everything from the heavy chernozem of the Holdrege-Hall in Kansas to the powdery whitish-gray solonchalk of the Lahontan Terminal in Nevada.

Although U. S. 40 follows an east-west line, it passes through an amazing cross section of climates, because of changes in altitude and rainfall. The roadway rises from sea-level to more than two miles above it, and passes close to peaks higher than 14,000 feet. Along the highway the annual precipitation varies from sixty to five inches according to distance from the oceans and relationship to mountain masses and altitudes. Steamboat Springs, Colorado, has a recorded temperature of fifty-four degrees below zero, but on the slopes of the Sierra Nevada foothills you see palms and orange trees flourishing.

Because of differences in climate the belts of natural vegetation also provide a cross section. In New Jersey the highway passes through coastal-plain woodlands of scrub pine. The hills of Maryland show all the richness of a mixed broadleaf forest. In Missouri oaks predominate. In Kansas the open grasslands take over, and beyond them you come to the coniferous forests of the mountains, and the pinyon and juniper of the plateau, and the sagebrush of the Great Basin, and the final ultimate barrenness of the salt-flats.

The fauna varies with the flora. With luck you will see a deer in Pennsylvania, and an antelope in Colorado. Even the pathetic casualties of the highway vary with the region. In the East it is usually a dog or a cat that lies sprawled where the car has tossed it; in the Rocky Mountains, a porcupine; in Nevada, a jack-rabbit....

U. S. 40 offers also a cross section in time. We need not fall back upon the anonymities of geology, or the vaguenesses of prehistory. We have good reason to believe, however, that in 1651 there was a short road, later called the "Street leading to ye woods," from Fort Casimir on the Delaware River, westward. This road may be taken as the beginning, in time, of U. S. 40.

Since 1651, a full three centuries, the route of U. S. 40 has been closely linked with the fortunes of the American people. It is the lineal successor, not only of the "Street leading to ye woods," but also of Nemacolin's Path, Washington's Road, Braddock's Road, the Frederick Turnpike, the Bank Road, the National Road, Zane's Trace, the Boonslick Trail, the Smoky Hill Trail, Berthoud's Road, the Hastings Cut-off, the California Trail, the Victory Highway.

Armies—in the French-and-Indian War, in the Revolution, in the Civil War—have marched and countermarched along it. Close beside it, young George Washington fought, and lost, his first battle. "The bombs bursting in air" that are commemorated in *The Star-Spangled Banner* must have been seen from points along it. In 1862 the battle of South Mountain was fought astraddle of it.

Besides Washington many others of our notables have been intimately connected with it. Jefferson, as president, interested himself in its development, and by his signature made it the first road to be improved at federal expense. Henry Clay supported its building, and scandal had it that the road was located through a certain city, where it still passes, because a certain lady made eyes at the great statesman. Farther west Daniel Boone is popularly credited with having opened up one section of the road, but this may be legend. It is no legend, however, that another famous frontiersman, Jim Bridger, helped explore the route west of Denver.

But U. S. 40 has figured in our history not only as a thoroughfare of war and as a field of action for our heroes, but also as one of the great routes of travel and commerce, east and west, along which have poured the anonymous thousands and millions, on their business or pleasure, on their vacations and their migrations, on foot or on horseback or by wheeled vehicle, through three centuries.

To adapt Voltaire's famous epigram: "U. S. 40 did not exist, and so the people of the United States—to fulfill their history—had to invent it!" To speak again in the person of the tour-conductor: "Yes, ladies and gentlemen, you may now observe, not only these United States as they now are, but also—if your sight can pierce that mist of time a little—you may see them as they have been in the past."

Nevertheless the same might be said of other roads, and the question may be fairly put: "Why U. S. 40, rather than some other?"

Actually, under the present national system of numbered routes, eight different ones may be considered transcontinentals. They are U. S. 20, 30, 40, 50, 60, 70, 80, and 6. (The famous U. S. 66 is not a transcontinental, but ends at Chicago.) Of these, 20, 60, and 70, degenerate badly in their western sectors. The others, however, are great highways and interesting routes, and about any one of them a good book could be written. Route 6, actually the longest of all, runs from the eastern tip of Massachusetts to the southwestern corner of California. Route 30, like 40, leaves one ocean at Atlantic City; it ends on the Pacific at the mouth of the Columbia River. Route 50 leaves Ocean City, Maryland, and ends with 40, in San Francisco. Route 80 provides a southern crossing, from Georgia, to southern California.

Perhaps the choice of U. S. 40 rests fundamentally upon the author's own whim. My destiny seems to have thrown me with that road. Once I wrote a book called *Ordeal by Hunger,* and found that to trace the course of the Donner Party I also had to follow long sections of U. S. 40. I found myself driving that same highway again and again when working on *Storm* and *Fire.*

Nevertheless, my own personal feelings discounted, a good case can be made for U. S. 40. It is not only, one can maintain, the richest historically of any of the transcontinentals, thus best demonstrating the cross section in time, but it is also a business-like modern highway, getting across from ocean to ocean with a minimum of deviation either for geographical or political reasons, and therefore displaying to best advantage the cross section in space. Its western terminus is only ninety miles south of its eastern terminus. If we take as its middle line a parallel running through Denver and Indianapolis, we find that the deviations, north and south, from this line are only 75 miles. Moreover, this route—by coincidence, crossing the 40th parallel four times—traverses the very center of the country.

By comparison, Route 6 runs uncertainly from nowhere to nowhere, scarcely to be followed from one end to the other, except by some devoted eccentric. U. S. 30 suffers because at its western end it slants off far to the north, and because in a long eastern sector it is rendered a secondary highway by paralleling the magnificent Pennsylvania Turnpike. U. S. 80, on the other hand, runs too far to the south, giving at most only a cross section of the southern United States. U. S. 50 is badly broken by Chesapeake Bay, and in sectors where it serves the same cities as U. S. 40—as between St. Louis and Kansas City and between Salt Lake City and San Francisco—it is the secondary road, carrying the lighter traffic.

Some people ask: "Why did you not choose the Lincoln Highway?" The answer here is easy: "The Lincoln Highway is dead!" The name, to be sure, survives in local usage. But the Lincoln Highway Association closed its offices in 1927, and most of its red-white-blue rectangles have long since been removed from the highways to make way for the white shields of the current national system. To retrace the route of the Lincoln Highway you would now follow chiefly U. S. 30 and 50, with shorter traverses over U. S. 1, 40, and other roads. . . .

To return to U. S. 40, one of its greatest attractions certainly lies in the direct way in which it crosses the continent along a central east-west line best showing variations of climate and vegetation, and at the same time displaying the topographical regions of mountain and plain, plateau and valley, that represent the basic structure of the continent itself.

Again we present our imaginary tour-conductor: "Yes, ladies and gentlemen, you are also seeing—on either side of the road—the continent of North America."

# HOW U.S. 40 BEGAN

On the evening of September 8, 1950, I sat in the Cosmos Club in Washington, and talked with Mr. E. W. James. Still actively at work, he was just back from Central America, where as Chief of the Inter-American Regional Office of the Bureau of Public Roads, he had just been inspecting the latest extensions of the Pan-American Highway.

"Just who," I asked, "is actually responsible for the route that U. S. 40 follows?"

With a pleased twinkle in his eye, he gestured backward with his right hand, pointing the thumb at his own sturdy chest. . . .

To appreciate the full situation we should go back a little in time. We need not, indeed, go clear back to 1651 or even to the days of Nemacolin's Path. All this more ancient history can come later, in connection with the different sections of the highway. For the moment, we need turn back only to 1912.

By that year the automobile had ceased to be an experiment, and was rapidly becoming the typically American means of transportation. Nevertheless, except under sunny skies, the automobile was marooned upon city streets. Most roads were abominable; in bad weather, impassable. Even where good roads had been constructed, they followed no real system. Main routes connecting large cities were almost unknown, because local pressures forced legislatures to scatter funds into the construction of short roads all through the state. As Carl G. Fisher wrote in 1912: "The highways of America are built chiefly of politics, whereas the proper material is crushed rock, or concrete."

In this year Mr. Fisher—most automobiles at that time used his Prest-O-Lite headlamps—conceived the idea of locating and constructing a transcontinental highway. He gathered behind the idea a small group of wealthy men. Shortly, they got a name, and organized themselves as the

Lincoln Highway Association. The leaders of the Association were deeply involved in the fast-rising automobile industry, and cannot have been unaware that improved conditions for automobiles might be reflected in increased dividends. Nevertheless, these individuals were not, in this enterprise, primarily concerned with profit. They themselves were automobile-drivers as well as automobile-manufacturers, and they liked to have good roads to drive on. Moreover, the project of a transcontinental highway was a grandiose one; it appealed to their imaginations, and allowed them to cast themselves as civic leaders and public benefactors.

Free of any noticeable taint of commercialism, consistently national in approach, the Association was remarkably successful. It adopted a standard road-marker, and patriotically appropriated the red, white, and blue for colors. Its activities impressed the name Lincoln Highway so strongly on public consciousness that one still hears it in ordinary conversation, even though the Association itself has not functioned actively for over twenty years. In order that the American motorist might have some roads that did not merely take him a mile out from the county-seat and then drop him into the mud, the Association explored and marked the first transcontinental road for automobiles, privately subscribed and collected large sums of money to aid in its construction, and turned the attention of state highway commissions and of legislatures toward this through highway, and thus to through highways in general.

But before long there were other and unforeseen results. Granted that the founders of the Association were large-minded and nationally oriented, yet each individual section of the highway was necessarily local. Soon restaurants and garages and hotels and whole towns were advertising themselves: "ON THE LINCOLN HIGHWAY." Obviously all towns could not be so located, but at least every town might be on *some* highway. Thus a whole new field suddenly lay open before that all too common American type, the promoter.

What you needed to do was merely to get one of the primitive road-maps of the day and trace out a route over some kind of road, good or bad, between one place and some other place. Then you gave this vague route a high-sounding name. You probably had an official marker sketched out on paper, but you did not need to get any of these manufactured. All you needed to do was to buy two cans of different colored paint.

You did not begin, however, even by painting telegraph poles. No, you followed the route, as best you could, in your own car. By good talking you could easily persuade Chambers of Commerce and individual merchants that it would be money in their pockets to have your highway come through their town, and that they should therefore subscribe to your association to see that the highway was so routed.

Actually, you may have been right; it may well have been worth their subscription to be located on the route. If, however, they remained deaf to your arguments, you had something more than words. You suggested that in such a case the highway could easily be routed so as to go through a rival town in the next county. If they were still unmoved, probably you actually did route the highway through the other town, even though that made a longer and harder route for the motorist who was following it. But what was still a better arrangement occurred when both towns were anxious to be on the highway. In this case you merely made a split, routed it both ways, and collected double.

With such methods and results, and with automobile travel becoming every year more profitable to garages and hotels and restaurants, routes multiplied amazingly. There were transcontinentals and north-south routes; others were fairly short and local. At least 250 are known to have existed.

Besides the name, each had a set of colors, to be painted upon telegraph poles, or fenceposts—or anything handy, to reassure the traveler. Thus, if you pulled up uncertainly at a muddy crossroads or fork, where there seemed to be only a choice of evils, you looked out anxiously, and seeing the proper set of colors on a pole, you turned in that direction. There was a Dixie Highway, using white-red-white, and running from the Great Lakes to Florida. Or you might follow the Old Spanish Trail, now largely taken over by U. S. 90 and U. S. 80; if so, you would look for a red, white, and yellow marker. The Cody-Billings Way used white-yellow-white; the Bankhead Highway, black-buff-black. It was not an era in which to be color-blind.

After a few years—by 1920, let us say—the situation was growing intolerable and ridiculous. In the first place, the contributions levied upon the local business-men went chiefly into publicity and salaries of the so-called "highway officials," not into good roads, or even to any great extent into good marking for roads. Moreover, the motorist was becoming tired of the situation. If he had a sense of direction, he would notice that sometimes the purple-and-red-painted telegraph-poles took him around three sides of a square, A-B-C-D, whereas he could just as easily and by just as good a road have gone from A to D direct. He realized that he had been bilked into going three miles to advance one mile because some merchant in the town had paid his contribution.

Besides, the "trails"—as they were commonly called—were getting too numerous. Sometimes the poles were painted almost from top to bottom. On certain roads as many as eleven different trails coincided. To spot the proper colors under such circumstances was something of a problem. Finally, you had to be careful which route of which route you were on. In order to collect as much money as possible some routes had two lines or even three. The Dixie Highway, for instance, had two branches over most of its extent, and had three branches north of Dayton and Indianapolis, plus a crossover between these cities.

Particularly irked were the officials of the various state highway departments and those of the U. S. Bureau of Public Roads. These men were responsible for the routing, construction, and maintenance of good highways, and for the wise expenditure of public money. But the linking together of these highways into interstate routes was largely in the hands of private promoters, and was being directed for private ends. The situation was much as if some profit-making organizations had grafted themselves upon the public-school system.

If the promoters had not been too grasping and had really worked for the public interest, they might have survived. But they went too far, and were too inefficient. By 1924 the American Association of State Highway Officials felt strong enough to begin action. The Secretary of Agriculture, under whom the Bureau of Public Roads operated, was asked to appoint a Joint Board on Interstate Highways, and this board reported at the meeting of the association in 1925. The secretary of the board, who read the report and was largely responsible for it, was Mr. James.

During its months of deliberation the Board had considered the question of the already marked "trails" and had decided to ignore them completely. This meant, unfortunately perhaps, the end of even such a national institution as the Lincoln Highway. The Board recommended a new country-wide system to be set up on a kind of checkerboard-grid. Routes of east-west trend were to be given even numbers; those of north-south, odd numbers. The chief east-west routes were assigned the numbers 10, 20, 30, etc.; the chief north-south routes, numbers ending either in 1 or 5.

The report of the Board was accepted, and in the next few years the system of numbered highways came to be universally accepted. A few of the old associations converted themselves, but their importance had vanished. They soon ceased to maintain road signs, and the bedizened telegraph-poles, in a few years, weathered back to gray.

U. S. 40, along with the other main numbered highways, thus originated in the mid-twenties from the work of the Joint Board on Interstate Highways. Since Mr. James was secretary of that board and acting chairman during much of the time, and since he assumes responsibility for the routing of U. S. 40, we may accept his explanation....

At the time of the Board's deliberations the chief transcontinental route was, beyond all doubt, the Lincoln Highway. Since it lay toward the northern half of the country, it received the number 30, in the eastern part of the United States. Perhaps the Board felt, however, that the complete taking-over of one of the older named highways in the new system would have resulted in pressure that other highways also should be taken over as a whole. In any case U. S. 30 ceased to follow the line of the Lincoln Highway in northern Utah and veered northwestward, along the line of the old Oregon Trail.

A little to the south of the Lincoln Highway an important route of 1924 —also using red-white-blue as its colors—led out of Washington, D. C., through Frederick, Maryland, and thus west to Columbus, Indianapolis, St. Louis, Kansas City and Los Angeles. It was known as the National Old Trails Association Highway, or merely, the National Old Trail. It served as the basis of U. S. 40 from Frederick to Kansas City. West of Kansas City, however, its swing-off toward the southwest made it ineligible for inclusion as a unit in the new system of east-west highways. Directly westward, however, other routes continued, at least on paper. By means of these the new U. S. 40 passed through Denver. By the time Kremmling, Colorado, was reached, all these routes had turned off in one direction or another except the so called Victory Highway with its yellow-blue marker. U. S. 40 followed this road on to the west, clear to San Francisco. Near Salt

Lake City the route of the Victory Highway and of the new U. S. 40 coincided with that of the Lincoln Highway for some hundred miles.

At its eastern end the new route departed from the line of the National Old Trail at Frederick, where that road swung south toward Washington. Continuing more directly eastward in accordance with the new east-west plan, U. S. 40 followed the old Frederick Turnpike to Baltimore. As Mr. James recalls, the original plan was to end the new highway there, at tidewater on Chesapeake Bay. It was, however, apparently in later meetings, extended eastward, first to connect with U. S. 13 in Delaware, and then clear to the ocean at Atlantic City. At the first official publication of the new routes, in April, 1927, U. S. 40 was a full-fledged transcontinental.

As thus established, it was essentially a combination of extensive parts of the National Old Trail and the Victory Highway, incorporating shorter sections of other roads. Like all the other routes laid out at that time it was a highway that hopped from city to city, not one that boldly leaped from coast to coast. Even as late as 1927, it is thus apparent, our most advanced highway designers accepted as their ideal a route that can be described as regional, but hardly as national. U. S. 40 usually offers the shortest and best route by which the motorist can go from Baltimore to Columbus, thence to Indianapolis, thence—city by city—to St. Louis, Kansas City, Denver, Salt Lake City, and San Francisco. But a motorist wishing to go by the easiest route from Baltimore direct to San Francisco would follow U. S. 40 only for a part of the way.

A man driving from Indianapolis to St. Louis and then to Kansas City could well follow U. S. 40. But if he wished to return directly to Indianapolis, he might better drive by way of U. S. 24 and 36. Thus the numbered transcontinental highways already begin to show something historical and quaint in their routes, looking back toward the time when there were few paved roads, and when even the most forward-looking planners thought in terms of only a few hundred miles between cities, not really of a transcontinental road.

The success of the numbered system was immediate and overwhelming—good evidence of how tired everyone was of the old "trails." Nevertheless, the new system did not escape all the evils of the old. Local pride and local business interests still objected to being left off the main road, and state highway departments were subject to such pressures. U. S. 40, for instance, split between 40 North and 40 South over a long stretch in Kansas and Colorado, and only in 1935 did the northern branch become U. S. 24. (U. S. 30 is still so split for several hundred miles.)

Moreover, when any highway was relocated, so as to by-pass some town or village, even the smallest community was likely to fight desperately to maintain itself on the through road. In many places the problem was solved by the comparatively innocuous method of establishing alternate routes, the more common arrangement being to call the old road the alternate. In some instances, this had strange results. An Alternate 40 appears on the map in eastern Ohio, even though it is only half a mile long and serves an inconsiderable village. Kansas, on the other hand, has solved the difficulty by tickling the vanity of some of the by-passed towns. Oakley (1159 pop.) and Sharon Springs (792 pop.) are both traversed by an official branch of the highway, marked *City Route*.

But all such charming anomalies, rather add to the interest of the route than subtract from it. U. S. 40 sprang from something that already existed, and has itself continued to change. It lacks, and will always lack, the logical and complete unity of anything that is created by a single impulse, such as a statue, or a dam, or a bridge. In its somewhat sinuous course from ocean to ocean, the highway sometimes swerves, for geographical reasons, because of a mountain or river. But it also sometimes swerves, for economic or political pressures, which are not less interesting because more subtle. Either of these latter forces may cease after a time to be operative, but the highway may still continue to run where it previously ran because of that force. Thus, in Nevada, U. S. 40 swings in a far northward loop, by way of Winnemucca, where it might run straight across from Battle Mountain to Lovelock. It follows its present course undoubtedly because the original emigrant-road went that way—because oxen and people needed to keep close to the river. In Pennsylvania a similar northward swing, by way of the town of Washington, still exists, though it was originally caused, a century and a half ago, by the rivalry of Philadelphia and Baltimore.

Much more about the history of the route can better be said in connection with the discussion of its various sectors. For the present we may turn to some general consideration of American roads.

# ROADS IN AMERICA

A nineteenth-century idea—startlingly simple and therefore attractive, as simplicity is always attractive—derived the routes of American travel from the original game-trails. An eminent and eloquent proponent of this theory was Senator Thomas Hart Benton, the great Missouri advocate of westward expansion. Speaking on the floor of the Senate, December 16, 1850, he declared in the best traditions of rhetoric:

> There is an idea become current of late—a new-born idea—that none but a man of science, bred in a school, can lay off a road. That is a mistake. There is a class of topographical engineers older than the schools, and more unerring than the mathematics. They are the wild animals—buffalo, elk, deer, antelope, bears, which traverse the forest, not by compass, but by an instinct which leads them always the right way—to the lowest passes in the mountains, the shallowest fords in the rivers, the richest pastures in the forest, the best salt springs, and the shortest practicable lines between remote points. They travel thousands of miles, have their annual migrations backwards and forwards, and never miss the best and shortest route. These are the first engineers to lay out a road in a new country; the Indians follow them, and hence a buffalo road becomes a war-path. The first white hunters follow the same trails in pursuing their game; and after that the buffalo road becomes the wagon road of the white man, and finally the macadamized or railroad of the scientific man.

Quite possibly the great senator was here making use of an idea supplied by, or strengthened by, his son-in-law, John C. Frémont. Known as the Pathfinder from his far-flung western explorations in the 1840's, the ever-energetic and sometimes muddle-headed Frémont had included in his reports an occasional note about a buffalo-path as broad as a road.

Even at that time, however, there were sceptics. One of these was Lieutenant George H. Derby, who as an officer of the Topographical Engineers and a western explorer, doubtless felt a natural pique at being put second, as a road-locator, to a buffalo or an elk. Writing over his pseudonym of John Phoenix—in a sketch which is really a parody of a Frémont report—he satirized, perhaps a little bitterly from his own experience: "the beautiful idea, originated by Col. Benton, that buffaloes and other wild animals are the pioneer engineers, and that subsequent explorations can discover no better roads than those selected by them."

Certainly the primitive country that was to become the United States was criss-crossed with game-trails, and in some places the buffaloes had beaten down broad paths. Certainly, also, there were well-established routes of Indian travel. Nevertheless, when we escape from orators and armchair-theorists and read the actual narratives of early explorers and road-openers, we find little mention of either game-trails or Indian paths. Possibly these explorers used earlier trails as a matter of course, and did not consider the fact worth mentioning. But this seems unlikely, and sometimes they even specify that there was no path at all, as did Augustine Herrman when crossing from New Castle to Elk River in 1659. So also—to use another example that will be later elaborated in a particular section of the history of U. S. 40—Berthoud mentions no game trail or Indian path to take him over the Continental Divide west of Denver, and apparently had to scramble up as best he could, finding the pass by trial and error.

Probably the primitive trails, in spite of their number, did not ordinarily run where the white men wanted to go. As Lowell Sumner, a wild-life expert, put it to me: "You can make use of an elk-trail for a while, and then all of a sudden it takes off—right up the mountain-side perhaps—some way where it doesn't make any sense for it to go."

As far as the origin of U. S. 40 is concerned, the buffalo-Indian theory of genesis remains almost wholly theoretical. Something is known about the origin of nearly every mile of the route, and there is nothing to indicate that game-trails or Indian paths were of importance. Even such an enthusiastic advocate of the general theory as Archer Butler Hulbert can only claim that Zane's Trace may have followed the "general alignment of the Mingo trail for a distance."

To pass from theory to history, one need only assert acceptance of the ordinary three-period division of American highway development.

The first period extends from the coming of the white men to the beginning of railroad-domination. During that period roads tended to extend and to improve. They began, in the forested East, typically, as narrow footpaths. Some of these footpaths can hardly have avoided following, in places, a game-trail or Indian path. Next pack-horses were taken over the trails, and where these large and somewhat clumsy animals, further impeded by their packs, could not follow in the footprints of their smaller and more agile masters, the route was altered or perhaps a little ax-work and spade-work was done. The two-wheeled ox-carts, next to follow after the pack-horses, demanded a still straighter and less hilly and more open course, and in the forested areas called for removal of more branches and underbrush. After the carts came the four-wheeled wagons and stage-coaches. These, being both longer and higher than the carts, demanded still straighter routes and yet more clearing of branches, and even the cutting of trees.

After the roads had been made passable for wagons, however, little improvement occurred in their surfacing, and because of heavier traffic many of them even deteriorated. Neither the finances nor the engineering of the time proved adequate.

Shortly before 1800, improved methods of road-building were imported from England and France, and an improved method of financing was developed in the form of the private turnpike company, which was empowered to construct and maintain the road and charge tolls for its use.

The first third of the nineteenth century was thus a period of steady, even rapid, development. Then, with the sudden success of the railways, everyone lost interest in roads, and even many of those that had been built were allowed to revert to gullies and mud.

The year 1835 is commonly given as the end of the early period of road-development. As with many other conceptions in our history this seems to be one which has been established with reference chiefly to Eastern conditions. Actually, in the West the period from 1835 to 1870 saw the greatest development in the opening of wagon-routes, and the establishment of the basic roads that have in the end become our national highways. After 1835 came the Oregon and California trails with all their numerous cut-offs and branches, and the many hundreds of miles of roads between mining-camps. Some of these, moreover, were well engineered and well constructed. Among such was the famous stage-road between Placerville and Virginia City, which still furnishes a route for U. S. 50.

The railroads did not begin to affect the West much until the completion of the first transcontinental in 1869. But after that date their effect was equally as striking as it had already been in the East.

The middle period of American roads—their subordination to the railroads, or their so-called Dark Ages—lasted thus from about 1835 in the East and from about 1870 in the West till the advent of the automobile, about 1900. During this period roads were conceived chiefly as local feeders for railroads, and farmers seemed to prefer hauling their produce to the railroad-siding over a bad road rather than laying out time or money to improve it.

The automobile changed the national attitude radically. After 1900 the interest in good roads increased rapidly, and the mileage of construction soared, decade after decade.

Viewed thus in the perspective of history, the situation and its development seem extremely simple. Actually, people living at the time, not blessed with clairvoyance, were startlingly unconscious of what was about to happen and was even beginning to be in the process of happening. Thus in 1899 the State Geological Survey of Maryland published an excellent work on the roads of that state and the possibilities for their development. The automobile is not mentioned.

Moreover, the commonplace that the automobile produced the good road, while it undoubtedly expresses the greater part of the truth, is not the whole truth. There is a certain chicken-egg relationship. Every new automobile created pressure for a new mile of improved road, but also every mile of improved road created the desire for another automobile. I can illustrate from my own memory.

In 1915 my father bought his first car. He bought it specifically because he frequently had to make a business-trip of about forty miles. Previously he had to make this trip by trolley and steam-train, but by 1915 the good road had been completed over the whole distance. The good road in this case, definitely produced a new purchaser, and therefore another automobile. There were certainly many thousands or even millions of Americans who, like my father, did not buy an automobile and then agitate for good roads over which to drive it, but waited until a good road had been constructed and then added to the traffic by buying an automobile.

William F. Gephart, in his interesting *Transportation and Industrial Development in the Middle West* (1909), gives contemporary evidence of the same state of mind: "When the improved road is a reality, an improvement in the motive power may be considered, but in the present state of highways it is useless to think of this." In the same passage the foresighted author goes on to predict the modern trucking highway.

But, though some might dimly foresee the future, no one—neither engineer nor statesman—could clearly foresee the magnitude of the future development, and plan a really adequate highway system. We may make a geological analogy. A river, having established a certain course, may maintain itself in that course even though it may have to cut out a deep canyon through a mountain-range that rears up through the ages. In the end the river may be flowing, paradoxically, through this canyon, when by a slight bend it might be flowing around the end of the range. But at no particular moment, throughout millions of years, was there a time when the river could actually have left its established course. In the same way a primitive road of 1900, itself developed from a packhorse trail, has been patched and relocated gradually, but the improvement has always been made on the basis of what was there already. Thus the pavement may have progressed from dirt to macadam to oil to thin concrete to thick concrete, and the route may have been straightened and made less steep, and the right-of-way may have been gradually widened. At length and after lamentable wastage of money, the path may have become a freeway, and yet the freeway may still bend here and there a little, because the path so bent.

We can distinguish again between the road as highway and the road as route. The highway of 1950 may have almost nothing in common with its predecessor of 1900. But the route of 1950, in spite of innumerable relocations, is often the obviously direct descendant of the route of 1900 or of 1800, or even of 1700.

In general the different sections of our modern highways represent very roughly four kinds of historical development.

Some of them—such as the sector of U.S. 40 in Delaware and eastern

Maryland—represent a full evolutionary sequence. These roads began as foot-trails or as primitive horse-trails, pushed on from one clearing or little settlement to the next. They became cart-roads and then wagon-roads, and perhaps turnpikes. They stood still or relapsed into the mud during the Dark Ages. With the coming of the automobile they were paved, and then gradually relocated.

Less commonly, a road was originally, in the early period of road development, laid out as a through highway. An unusually large part of U. S. 40 is of this nature, since it follows, from Maryland to Illinois, the line of the old National Road, which from its very beginning was planned as a through route, without reference to previously existing roads.

In the third place, during the period of railroad-domination, many roads began by paralleling the railroads, particularly in the Plains states. Thus, across most of Kansas and eastern Colorado, U. S. 40 runs close to the Union Pacific. Roads tended to follow a railroad because the railroad itself was often the shortest line between two towns and because the towns were strung along the track. Moreover, the right-of-way for a road could be established alongside the railroad with a minimal disturbance to farms and private holdings. Also, the roads may have followed the railroads for safety. If you bogged down hopelessly in the gumbo or broke an axle, a freight-train or a handcar might be prevailed upon to stop and take the wife and children into the next town.

Finally there are a few modern highways which are wholly recent in design and also in route. U. S. 40 in Nevada, from Oasis to Wells, follows neither railroad nor emigrant trail, but seems to be wholly of the automobile period both in construction and in route.

In my own mind I find that I can also classify highways advantageously as dominating, equal, or dominated. A dominating highway is one from which, as you drive along it, you are more conscious of the highway than of the country through which you are passing. Six-lane highways, and four-lane highways, particularly in flat country, give this impression. You see the highway itself, the traffic upon it, and the life that has grown up along it and is dependent upon it—all the world of service-stations and garages and restaurants and motor-courts.

To many people, of whom I am one, parkways produce the same effect. Although esthetically beautiful, the artificial landscape on both sides of the parkway becomes part of the road itself, and is divorced from the countryside and from reality. The parkway by-passes towns, and therefore the motorist has no sense of actuality. A parkway is excellent at providing unimpeded transportation, and for allowing the city-dweller his escape, but when you drive along the parkway, you are not seeing the real United States of America.

The dominated highway, on the contrary, is one which seems to be oppressed and to lose its own identity because of the surroundings through which it is passing. Highways are dominated when they pass along city streets. There is too much close by on either hand. There is too much local traffic that has not the slightest concern with the farther reaches of the highway. On the other hand, highways may be dominated when they are comparatively small roads passing through high mountains or vast plains. Again the highway becomes insignificant, and one's interest is pulled outward, away from it.

In between, lies the equal highway, that one which seems to be an intimate and integral part of the countryside through which it is passing. On such a road there is a division of interest between one's focus upon the highway and its margin and upon the country back from the highway. . . .

A road has been called a "symbol of flow"—not only of people and things but also of ideas. Close the roads, and you block the flow of ideas. Thus the Iron Curtain went down across the roads of Europe, and all the modern devices of printing and radio have not been able to compensate.

At the present time we must take it as a matter of pride that the motorist can drive the whole length of U. S. 40, the breadth of a continent, with no more impediments to his progress than traffic-lights, a few toll-bridges, and the inspection for plant parasites at the California border. This last is the only one to which objections can well be taken. If a plant quarantine is necessary—and it may well be—the administration should be in federal hands. If other states should insist upon inspecting every citizen's personal baggage when he crossed its line, the infringement upon personal liberties would become intolerable. Such a situation would be a symbol that our country had taken a long step backward, toward disintegration. Fortunately the present situation along the length of U. S. 40 gives no indication that we are moving in this direction, and the motorist ordinarily flashes by a state line without even realizing that he has crossed it.

# ON MOTORING

Considering the fixation of the American people upon automobiles and their use, there has been surprisingly little of what might be called personal or imaginative writing on the subject. In the first decade of the century C. N. and A. M. Williamson delighted the reading public with a series of novels under such titles as *The Motor Maid,* and *The Car of Destiny.* But since that time has there been a novel based chiefly on motoring?

One can of course think of sections within various novels. There is, outstandingly, the journey of the Joads in the *Grapes of Wrath.* I remember also the flight and pursuit across Montana in Stegner's *Big Rock Candy Mountain,* and the rollicky ride of the college boys across the Middle West in Feikema's *Primitive.* If you will pardon a personal reference, there is even a good deal about automobiles in my own *Storm.* Poets, I think, have done very little with the subject, although I remember a few good lines by Josephine Miles about a nervous passenger, and some of us can remember Vachel Lindsay chanting his *Santa Fe Trail.*

The social historians, of course, have written about the automobile as a moral, or immoral, influence. These masters of statistics, however, seldom get much under the skin of the individual.

But, if I look back over my own life—being at the moment fifty-six years old and still of sound memory—I find rather few things that have meant as much to me as automobiles and motoring. I think that I could even borrow a title formula from Betty MacDonald, and write a book called *The Automobile and I.* At least, I might here and now write a few such paragraphs....

I was born in 1895. Just when the automobile was born may be considered a matter of controversy, but for our purposes we may use the year 1893, when Duryea first successfully operated an automobile in the United States and Henry Ford built his first car. The automobile and I are thus very close to being exact contemporaries.

For some years I remained quite unconcerned with the automobile, and it was undoubtedly quite as unconcerned with me. As early as I can remember, however, I liked to ride in a buggy, and so I was obviously developing into an easy mark for the automobile salesman.

When I first heard of such a thing as a horseless carriage I can't remember, but I remember the first time I ever saw one. We lived in a small town in western Pennsylvania, and on one occasion my father took me on a trip to Pittsburgh. I don't remember what the year was; it might have been 1902. As Thurber's character says, "You could look it up," for I remember that *Buster Brown* was playing at a theater, and my father took me to see it. It was the first time I had ever seen a play, and I can still remember some of the lines. I must have been a very small boy, for on the way home from the theater I remember looking up and seeing some of the lights at the top of a distant skyscraper, and asking my father if those were stars.

No matter what year it was—across the street filled with cabs and other horse-drawn vehicles, I remember, we saw a strange thing without horses. We went over and looked at it. Not to keep anyone in suspense, I shall state immediately that it was an automobile. At least, it was what passed for an automobile in those days.

I remember also the first automobile that came to our small town. Again, I don't remember the year, but it might have been 1903. I would have been eight years old then, and I can remember that all the boys were much excited with the idea that we were going to have an automobile in town. The morning it was scheduled to arrive, my father and I happened to drive a little way out of town in a buggy, and he pointed out to me in the dust of the road a broad mark, broader than would be made by the ordinary wagon-wheel. That, he said, was the mark of the new automobile.

When my uncle actually bought the second car to come to town, I was much set up. It was one of the kind that was widely advertised as having been able to climb up the steps of the national Capitol. It had two seats, and you climbed into the rear seat by a door that opened at the back. After a while I actually had a ride in this car. I can state positively, I believe, that this first ride was in the year 1904.

Well, the automobile and I moved toward adolescence together. I became one of the generation of boys who sang "In my merry Oldsmobile," and shouted "Get a horse!" Automobiles became fairly common, even in a small town.

In 1908 my family moved to southern California, and there roads were vastly better, and automobiles correspondingly commoner. There were even a few trucks.

My family did not have a car, and I did not learn to drive until 1914, when I was nineteen years old. These days, that sounds a little like learning to drive in one's old age, but it was probably average for the time. As was just right to happen, I learned on a Model-T, and a not very new one. I happened to be in Pennsylvania again that summer, and I learned to drive a good and hard way, on the steep and crooked and muddy roads of Westmoreland and Indiana counties. Learning to drive with that car on those roads was a fine experience, and since then nothing about driving has ever bothered me particularly.

But we have time only for the high spots. In 1915, as I have recorded already in another connection, my father got his first car, also a Model-T, and I taught both my father and mother to drive, something that very few children in the future will be able to say. Also, many years later, I taught my daughter and son how to drive; my wife, I am glad to say, came to me fully competent.

In 1917 I served in the United States Army Ambulance Force, and there too I drove a Model-T. Actually I did not get overseas or drive an ambulance very much, but it is at least another point of contact.

In 1923, I bought my own first automobile—a second-hand touring-car. Since that time I have owned—incredible though it seems to me, as I count them up—either personally or as head of the family, eleven automobiles. This is all the more remarkable since for a number of those years I owned no automobile at all. To list them in no particular order, they were one Oldsmobile, one Buick, two Chevrolets, two Overlands, two Studebakers, and three Fords. In type they have been one convertible, one station-wagon, one jeep, one two-door sedan, one four-door sedan, three touring-cars, and three coupés. Five of them I bought as new cars; six, at second-hand.

In these cars I have driven widely, in every state of the union and in Alaska and the District of Columbia. I have also driven in twenty Mexican states and five Canadian provinces. In latitude I have driven from Fairbanks in Alaska to Acapulco in Mexico; in altitude, from Bad Water, California, which is more than a hundred feet below sea level, to the crater of the Nevado de Toluca which is about fifteen thousand feet above.

I mention all this only to illustrate my relationship to the automobile and as something from which I have had a great deal of pleasure, and not as anything which I think to be particularly unusual. There are many hundreds and thousands of Americans who have driven as far as I have, have owned more and better cars, and have been much more closely identified with automobiles. Although I have changed many tires and have on occasion cleaned spark-plugs and tightened brakes, I am not one of those who really enjoys going into the insides of an engine and knows much about how to do it. The fastest I have ever driven a car is, I think, 83 miles an hour. The longest run I have ever made in one day was from Salt Lake City to Berkeley, California, a mere matter of some seven hundred miles. Although I have driven quite a few thousand miles, I am very far from setting any record for amateurs, and an attempt to estimate my number of miles would probably result in a disappointing figure.

On one detail I am proud, although I also realize that I have been lucky. I once turned a car on its side in the soft ground of an orange grove, and scratched my wife's wrist. Otherwise, I have never hurt anybody. I have not even damaged a car seriously. (Of course a great many things can happen to me any day, even before this can be published.)

I have given some thought to the question of why I enjoy motoring, and why other people also enjoy it. One can of course fall back upon the old saying that Americans, from frontier times downward, have been a foot-loose people, always moving on. But for the vast majority of people motoring is not a moving on, but merely a departure from a fixed point and a return to that point.

One thing that must be recognized, however, is that motoring must be, for most people, an end in itself, not a means of getting somewhere. People will drive for hours, even days, across uninteresting country, only to stand for a few minutes to "Oh!" and "Ah!" at some waterfall or chasm. If they did not really enjoy the mere motoring, it is incredible that they would drive so far to spend so little time looking at what they went to see.

There are of course many different kinds of motorists, but I believe that a great deal of this charm for most people can be placed under two heads.

In the first place, many people like to drive, even on city streets, in order to satisfy their longings for power. In a car even a physical weakling can go far and fast, and he can make pedestrians, and even most other cars, get out of his road. This satisfies his need for aggression, which is tied up with his desire for power, and thus he is basically happy when driving. This type of driver is more commonly a man, but some women react to the same urges.

In the second place, people like to drive because driving is, actually and symbolically, an almost perfect mechanism for escape. This is, I think, my own chief source of pleasure in it. There is probably no human being who does not have troubles, real or imagined, from which he at times feels the need to flee. People find this means of escape in many ways—by playing golf, or canasta, or by going to the pictures, or reading books, or even writing books. For myself I find no better means of escape than to get into a car and start driving, particularly if I am not that same night coming back to sleep at the same place. I have often actually felt myself relax and feel contented, as the road began to flow and I looked ahead with no more immediate concern than the pleasant matter of steering around the coming curve, and began to sense the continual moving toward, with the necessarily concomitant moving away from.

By association there has grown up a whole group of pleasures, even of primitive sensuous pleasures, that are a part of motoring.

There are the visual sensations. . . . The low sun behind, with the shadows long ahead. The low sun ahead with the shadows pointing this way, and the glare in the eyes. The sun high above, and the light full on the road, with the shimmer of heat-waves rising from the asphalt. No sun at all and the darkness thick, with the beams of the headlights picking out the curve ahead.

There are the sounds. . . . The continual low muttering talk of the engine. The whir of the tires on concrete, and the shift in tone that comes with asphalt surface, and the still louder whine of brick. The soft squish of wet tires on pavement running with rain. The continual whirl and hiss of air around windows. The little click of windshield-wipers. The quick *zee-ip, zee-ip* as you pass something close by, a sound that in childhood was associated only with railway trains.

There are the smells. . . . The freshness of early morning before the day has matured, and that other freshness that comes after the thunderstorm. Now and then, the gasoline odor, or the diesel odor, as a truck goes by. And the smell of the car itself too—as comforting as home, as familiar as old shoes—a subtle blending, of oil and upholstery, and cigarette smoke, and people.

There are, too, those vague sensations that we call kinesthetic. . . . All the rivers of fresh air coursing over the face. The pressure backward with acceleration, and the pressure as the body swings forward when the brake goes on. And the continual joggling from the springs, doubtless good for the digestion and the nerves and the general well-being, reminiscent perhaps even of the joggling of the child within the womb.

All these are familiar to us Americans, and they will be, we hope, familiar to the Americans who come after us. But only the generation of which I am one will ever have grown up along with the automobile, and have experienced the miraculous unfolding.

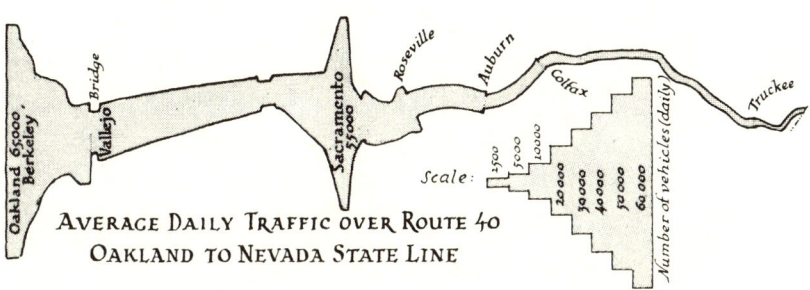

AVERAGE DAILY TRAFFIC OVER ROUTE 40
OAKLAND TO NEVADA STATE LINE

**THE PICTURES**

## AS FOR THIS BOOK—

An author can exercise no authority over the way in which his book shall be read, and even the offering of a suggestion is perhaps an impertinence. What I hope, however, is that the reader, in the sections presenting the pictures, will neither go through merely looking at the pictures nor merely reading the text, but will read a sentence or a paragraph and then look again at the picture, actually to see what the text has pointed out as existing there, and thus constantly shift from one to the other, to fit each picture and each part of each picture into its place in the cross section of the United States.

Several years ago the idea of a "picture-book" began to intrigue me, and I considered methods by which text and pictures could be better coordinated than they were in books that I had seen. In this book I have tried to attain this coordination by a method that I shall explain shortly.

I am sometimes asked: "Why did you not team up with a professional photographer?" The answer is, partly, that I like to take pictures. But also there were serious practical difficulties in the way.

In my pictures I have tried not to enter into competition with either the "esthetic" or the "news-value" school of photography. I have tried to take pictures which might have some esthetic value, some "total effect," even some "center of interest," but which might also be conceived as a collection of parts, besides being a single whole. You may, if you wish, look at each picture as a unit, but you should also look at it as a foreground, a middle-ground, and a background, as a right, a center, and a left. Moreover, I myself have studied every picture carefully and have attempted to learn what it shows. In the accompanying text I tried to show the viewer what he can derive from the picture, in addition to a possible esthetic feeling. In a sense I have returned to a nineteenth-century theory of art by which: "Every picture tells a story."

Obviously my theory is realistic. I am attempting to offer a portrait of U. S. 40 and of the accompanying cross section of the country. Like Walt Whitman, I reject nothing. A friend looked at one of the photographs and remarked "Very good! But it's too bad those wires came where they did." He was obviously speaking as one of the esthetic school. I explained to him that far from trying to avoid the wires I had maneuvered myself into a position where the wires were emphasized. They are certainly part of the United States and what I was trying to do in that picture was to indicate the way in which all forms of transportation, including power-lines, bottle-neck along with the highway in traversing a mountain pass.

Since I am striving to present the cross section, I have been interested in the typical as much as in the outstanding. I have sometimes even avoided the picturesque. Also, since I was necessarily dealing with the present in the pictures, I have treated the history in the introductory sections, and in connection with the pictures have generally dealt with what can be seen. Only a few, such as those of the restored Fort Necessity and of Braddock's grave, primarily recall the past, and demand a largely historical exposition.

I have not—and I have been very conscious about this—paid much attention to cities and to people. The American city, east and west, is highly standardized, and a picture of one city stands pretty well for a picture of another. Much the same may be said about our citizens. They show plenty of variety indeed, but this variety is not a reflection of their east-west location on the highway nearly so much as of such other matters as racial background, and economic status. For instance, you will see truck-drivers all along U. S. 40. Some of them are big; some, small. Some are smiling-faced; some, hard-faced. But there is no essential difference according to region. The landscape will vary tremendously, but whether the background shows the Jersey coastal plain or the Rockies of Colorado, the driver—and his truck too—will not differ except by the mere accidents of individual variation.

Much, also, I have had to pass over because of limitations of space. I have prepared a single and somewhat impressionistic work, not a complete guidebook or an encyclopedia. Inevitably one picture has come into competition with another, and I have had to make a choice. Sometimes a scene with interesting historical background yielded only a dull picture. This happened, for instance, with the prehistoric Cahokia mounds past which U. S. 40 runs just east of East St. Louis. Sometimes, on the other hand, I rejected a fine picture because there was little of interest to be written about it, perhaps because it to some extent duplicated the effect of some other picture.

I have also tried to keep the pictures from dominating, as they do in most "picture books," and at the same time to keep the text from dominating, as it does in books which are merely "illustrated."

Primarily and always, my subject has been the highway. Sometimes it holds the foreground; sometimes its pavement is far off in the background. It is always there somewhere. Yet I have tried to photograph it, not as a thing in itself, but as a part of its setting, and often the setting is the more important. Except in California and Nevada the pictures were taken during trips that I made along the whole highway, in May and June, 1949, and in August and September, 1950. In the two westernmost states I was able to take several shorter trips, at different seasons of the year. Theoretically I should have taken at least four trips, to display the cross section at different seasons. This was impractical. In any case, I repeat, this book is not an attempt to produce an encyclopedic work.

Since U. S. 40 is very much contemporary—not, like the National Road, a part of the unaltering past—this book necessarily presents the actuality only of a particular time. Even between my first trip and my second there were noticeable changes—alterations of grade and pavement, openings of new alternates, whole relocations. In 1950 the motorist still had to take the ferry between New Castle and Pennsville but the Delaware Memorial Bridge is now open, and wheels can roll all the way from coast to coast. In describing some of the pictures I have mentioned the imminence of change, but a continual repetition would become monotonous, and on the whole it seems better merely to write a general warning: "Thus it was when I passed by, in my time."

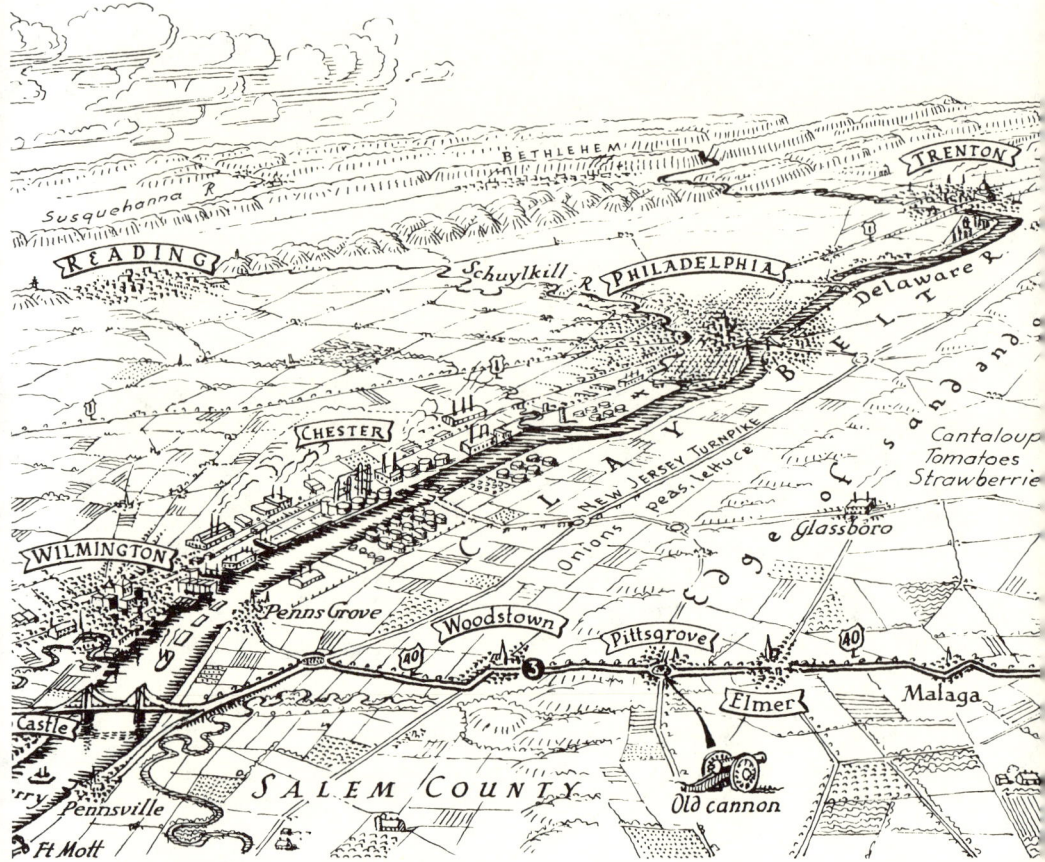

## JERSEY PROLOGUE
## Atlantic City to New Castle

Normally, a U. S. numbered highway, like a rope, is considered as having two ends, and no beginning. As far as I have been able to ascertain, however, U.S. 40 is abnormal in having only one end, at least only one that is officially proclaimed by a marker, and that one is in San Francisco. Atlantic City, apparently, has never taken its eyes off its bathing-beauties long

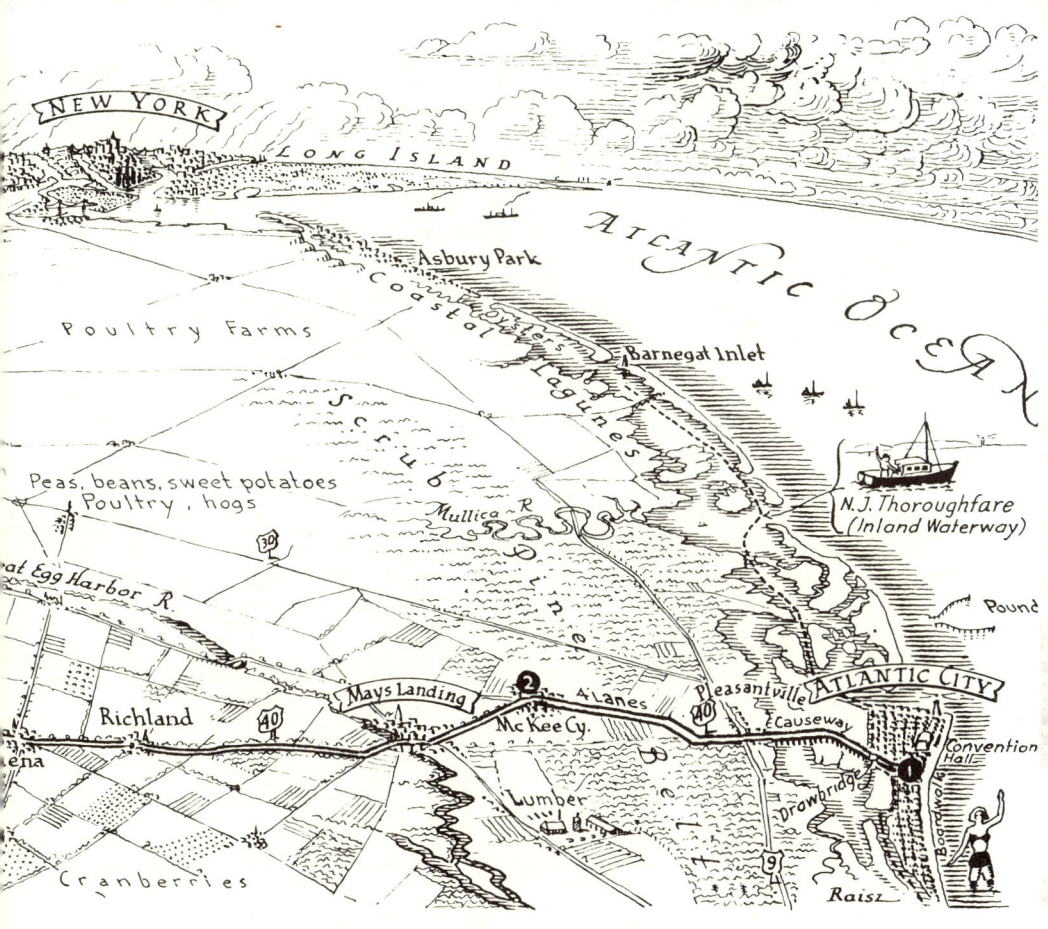

enough to establish a terminus—but let that wait till we get to the picture.

Since in some sense, however, U. S. 40 must have two ends, how shall we decide which one shall function as a beginning for the purposes of this book. For, obviously, though a highway can be traveled in either direction, a book can be read only in one. Thus forced to a decision, I have chosen to begin in the East and go west. This is a choice which is based perhaps, pusillanimously, upon analogy and convention. East-to-west is the course of the sun, of (it has often been asserted) the human race and the course of empire, of (certainly) the historical development of the United States. Therefore, most of our geographical writings follow the pattern, and our atlases (unless they seek refuge in alphabetical order) present the eastern seaboard first.

An interesting effect might indeed have been attained, in this book, by starting in the West and by coming east, thus gradually piercing down through layer after layer of our history. The final difficulty, however, would have been that by so doing the book would have ended at Atlantic City—and to cross a continent and end on the boardwalk seems anticlimactic, not to say downright ridiculous. But, in any case, we begin....

The W.P.A. Guidebook for New Jersey, a curiously jaundiced work, describes U.S. 40 across its state as a "bog-to-bog" highway. Although literally correct, since the road passes from tide-flats along the Atlantic shore to tide-flats along the Delaware estuary, this description is in effect a slander. For my money, you will find few pleasanter little runs than the 68 miles westward from Atlantic City.

Spectacular it is not—most certainly! It follows a causeway across the salt-marshes of the coast, goes through the scrub-pine belt, rises a few feet into a country of fields and pastures and woodlots, dips down again to tide-level. Pittsgrove, the highest town on the route, is only 153 feet above the sea, and the loftiest hill to be seen is merely some inconsiderable wrinkle of the Coastal Plain.

For the first thirteen miles, to McKee City, there is a four-lane highway, but from there a delightful little two-lane road leads on westward, not heavily traveled, having few straightaways for such flat country, and curving intimately among almost imaginary hills and through quiet towns and villages.

Of the excitements of history this region has known little. New Jersey was not among the first colonies to be settled, and this southern end, much of it even yet forested, did not attract colonial farmers. It long remained, and much of it even still remains, backwoodsy—rural at most.

Even Atlantic City, the terminus, far from being colonial, is a more recent settlement than Kansas City or Salt Lake City, and having been from the first an amusement-center, it represents the froth rather than the sound wine of American history.

Some of the villages west of the pine-belt saw a few red-coated foragers during the Revolution. A cannon, of uncertain history, dated 1763, still looks out over the peaceful traffic-circle at Pittsgrove. A few eighteenth century-houses stand in the towns and along the highway.

The route itself, like the region, offers little of historical interest, consisting only of village-to-village roads strung together, and never having been an important line of travel. Like so many roads, it is said to follow an Indian trail, and quite possibly it does in part—for in these eastern woodlands Indian trails were plentiful, and the land was criss-crossed with them. But since in any case there were no natural barriers, the existence of an Indian trail cannot have been of very great importance in aiding the establishment of a road. Around Woodstown and Pittsgrove the route is still called locally the Pole Tavern Road, because it once led to that well-known tavern. During the "trails" era, this trans-Jersey route was incorporated into one of the late-comers as part of the Harding Highway, a road that seems to have met with no more success than the president for whom it was named. It has vanished with so little trace that I have not even been able to discover its colors.

The short New Jersey sector is a pleasant prologue, but nothing more. As a great modern highway and as a historic route, U.S. 40 really begins on the west bank of the Delaware River.

# ❶ Beginnings of U.S. 40

US 40 In keeping with its utilitarian and non-ceremonial nature, U. S. 40 has no monument or even signpost to mark its beginning. It merely emerges from the streets of Atlantic City. This is unusual. Most federal highways come to a definitely and clearly marked terminus, but in Atlantic City the westbound motorist must merely, so to speak, pick up the trail by sighting a seemingly chance marker along the street.

As a practical starting-point, we can take the traffic-circle surrounding the World War Memorial. From this point westward, for nearly three thousand miles, the route is posted continuously with the shield-shaped markers of the federal highway system. Here at the traffic circle U. S. 40 is only a quarter mile from the ocean, though actual view of it is cut off by the intervening Atlantic City board-walk.

As seen in the picture, taken from the roof the President Hotel, U.S. 40 begins with a mile-long reach. Though going directly away from the ocean and starting to cross the continent, this first stretch runs only a little west of north—for the highway's first task is to get across the difficult marsh-and-lagoon belt. At the end of this initial straightaway the highway swings westward to cross the marshes on a slightly raised causeway. In the distance its course is outlined by the billboards lining it.

Topographically, the picture displays the Coastal Plain. Tipped almost imperceptibly toward the east it lies awash—doubtful, one might think, as to whether it really wants to be land or water, or whether a choice is actually necessary, and not making up its mind until after using up five miles of no-man's-land consisting of marshes and tidal channels.

Atlantic City stands on a bar built up by wave action, and inland there is, as the picture indicates, more water than land. Beyond these marshes and lagoons the "old land" of the Coastal Plain rises slightly, marked in the picture by faint white spots, indicating the towns built along its edge, above the level of the marshes. But not the slightest bump appears on the sky-line; not even the horizon of the High Plains is leveler.

The nearest channel in the picture is that known as the Inside Thoroughfare, and forms a part of the inland waterway by which small boats can pass along most of the Atlantic Coast, largely by utilizing such behind-the-bar channels as this. The highway crosses it by means of a drawbridge, the first of some hundreds of bridges along the route.

The outlying district of Atlantic City here in view shows the characteristic American attempt to maintain gentility by means of a detached house, even though the houses have to be packed tight together. As in all cities built on small islands, space is at a premium.

In the background can be seen those common appurtenances of an American city, an athletic field and an airport.

The only building to have architectural pretensions is the War Memorial itself, an open circular cella with surrounding peristyle of Doric columns. Through one of its four doors may be seen the dark form of the heroic bronze Liberty-in-distress, about which the New Jersey W.P.A. Guide comments with characteristic acidity: "Her distress is the one matter about which there is no question whatsoever."

# ❷ Coastal Plain

**(US 40)** As seen from the McKee City firetower, the highway runs a little south of east toward the ocean across the plain. In a sense, this is the reverse of the previous picture. Under the near-monotone of gray sky the towers of Atlantic City break the distant horizon slightly, beneath a faint smudge of smoke. One might say that the Atlantic Ocean itself is in view beyond them, at least the air above it.

U. S. 40 along a straightaway of some five miles here presents the aspect of a fine modern highway. It consists of two adequate lanes for travel in each direction, separated by a well-kept safety-zone with numerous crossovers. Outside the lanes of travel are broad shoulders for slowing down and parking, with good ditches for drainage, and pole-lines set back safely outside the ditches. The dark bituminous-surfaced east-bound roadway is the older, and contrasts with the still largely white concrete surface of its counterpart. The oil-stains on the concrete lanes, much heavier on the outside lane, give evidence that most drivers habitually keep out of the inside lane except when passing. This conclusion contradicts the actual traffic-situation of the moment, for about as many cars can be counted in the inside as in the outside lane.

The picture also demonstrates the diseases of a modern highway. The numerous spots of white paper on the dividing-strip mark the accumulating debris thrown out by careless travelers. Billboards have been erected among the trees. Already an enterprising advocate of free enterprise has established some overnight cabins, and their laundry is hanging out to dry. The sheets and towels may be accepted as a mere homey touch, but if more cabins are built, the free flow of traffic will soon be impeded, and a reduced-speed zone will become necessary. Already the need of a freeway, or limited-access highway, is becoming apparent.

As in the picture taken from Atlantic City, the landscape represents the almost incredible flatness of the Atlantic Coastal Plain. Although thirteen miles from the ocean, the road at the foreground of the picture is only about seventy feet above sea-level.

The vegetation is scrub pine, here representing the most northerly wedge of the great pine-belt that sweeps up the Atlantic Coast clear from Georgia. Because of the thin soil, farming does not pay in this area, and it will probably remain permanently forested. Although trees grow vigorously, they remain individually small, and the forest is of secondary quality.

# ③ Farmhouse

🛣️40 Near Woodstown, facing the highway from the north, stands a good early American house. Mrs. Annie Newell of the Salem County Historical Society, who has kindly investigated for me, has been unable to determine its date: "The house is old, the taller part of the house. How old, no one seems to know." She has been able to determine only that it was built before 1865, but from its general appearance, it may well date back to about 1800. Mrs. Newell is understandably a little put out with me for having selected this house instead of one of the eighteenth-century brick houses for which Salem County is famous. None of these, however, stands in photographable position with respect to the highway.

Nevertheless this house is far from being without interest, and in its general appearance is typical of many old New Jersey houses. It is constructed of brick, made on the place. More recently it was cemented over on the outside. The windows in the end, indicate that it was built at a time when people were thinking of farm houses, not of houses jammed into a row along a street. Even so, in its rigid compactness and lack of adjustment to outdoor living the house bespeaks the townmindedness of most of the

English colonists. Even if they had been farmers, they had generally lived in close-built villages, and they took several generations to adjust themselves to living on the isolated farms of the colonies. The porch here was actually built at a considerably later date than the house.

The windows are double-hung, a typical feature of American architecture, highly functional. Except for the two little windows in the garret, all windows are well shuttered, and they are screened against the insect pests of the long and hot summer. Although the trim whiteness is suggestive of New England, the off-center door is much more a feature of the middle colonies.

Another American feature is supplied by the prominent lightning rods, on both house and barn, probably a contribution of the lightning-rod-salesmen era of the later nineteenth century.

The house is somewhat undersupplied with chimneys, and probably most of the cold-weather living has centered around the kitchen stove. But the winters of southern New Jersey are comparatively mild and short, so that the tremendous fireplaces of the more northern states are not so much needed.

The structure shows little adaptation to the summers. The fine willow tree stands at the northeast corner, where it would scarcely be of use for shade. The eaves are skimpy, permitting even the noonday sun to have full play upon the windows.

The barn is also a good piece of Americana. It shows the overhang by means of which hay can be hoisted up into the loft. It also shows the advertisement for chewing tobacco, which is so common in many parts of the country as almost to be conventional. The cornfield across from the house is an equally sure American touch.

At this point the highway crosses the gentle crest of one of the inconsiderable rolls of land displayed by the Coastal Plain. The earlier road crossed the natural crest. The present highway, as shown by the banks on either side, has been cut down by about five feet.

At the left a local telephone line of ten wires follows the highway. The two poles at the angle, where they are subjected to sideways strain, have been guyed into the bank for additional support.

At the right a local power line, its pole at the angle reenforced by a push-strut, sends off an array of wires both to the house and to the barn.

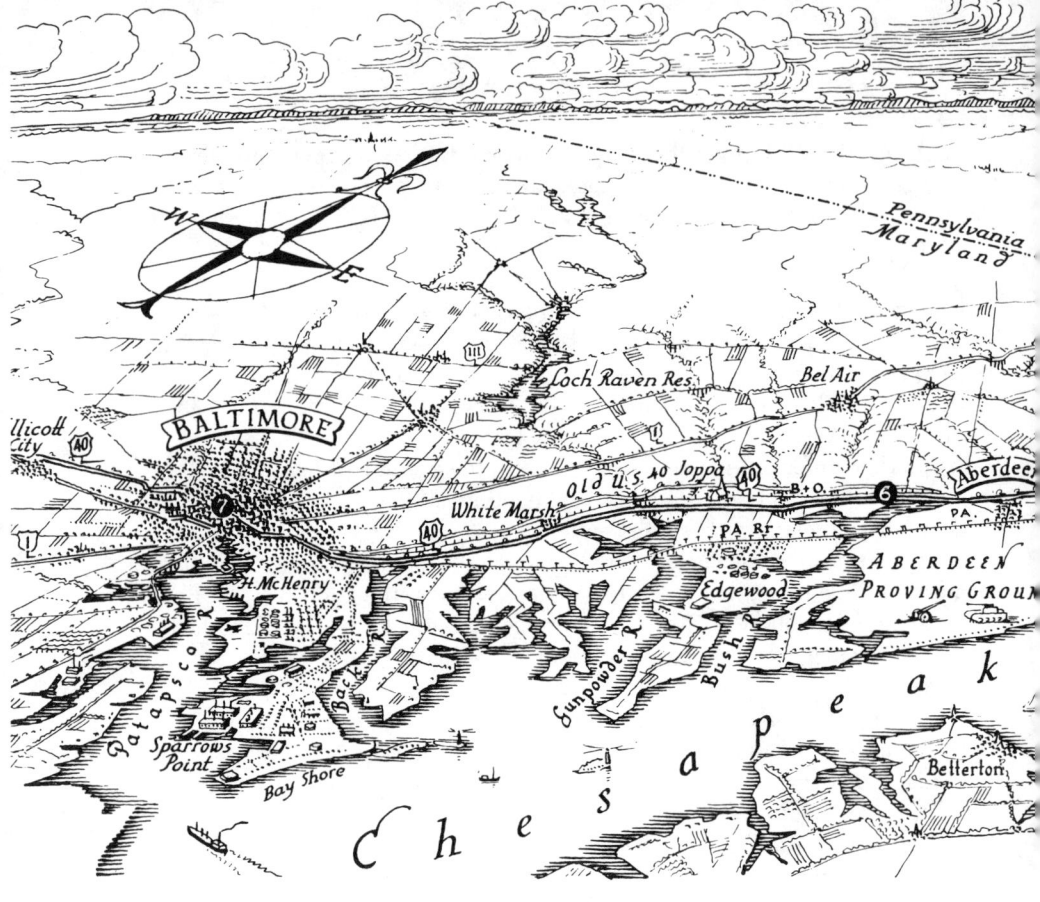

## POST ROAD

### New Castle to Baltimore

U.S. 40 in Delaware and eastern Maryland, considered merely from the point of view of what can be seen along it, is as dull as any section of the whole route. It is, however, the oldest part of the whole, and in historical interest the equal of any, though the total distance is only 65 miles.

Topographically, the road here follows closely along the boundary between the Atlantic Plain and the gently rolling Piedmont, but the general

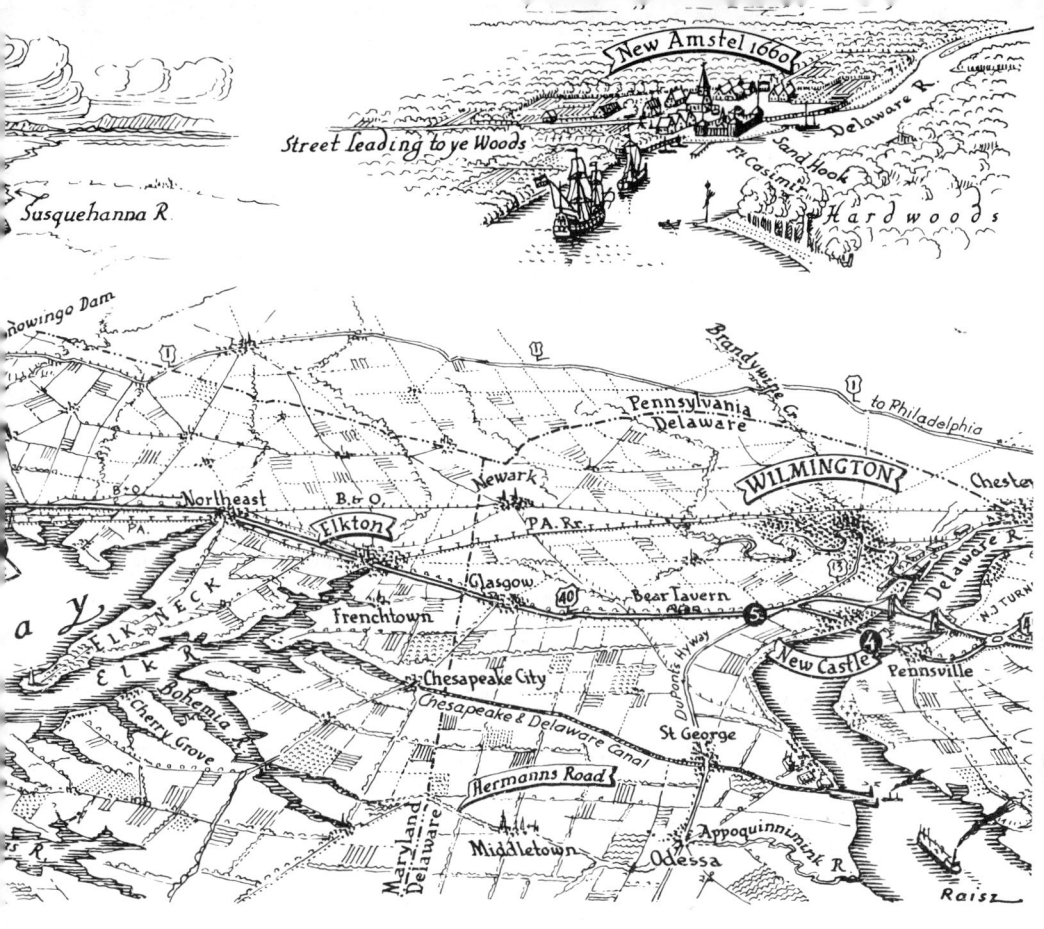

impression is one of flatness, without the vastness that lends majesty to the Great Plains. Much of the land has gone back to scrubby woods, and as elsewhere along the Atlantic coast, the traveler feels the anomaly between busy modern cities and a run-down, slatternly countryside.

The highway itself dominates. With the towns by-passed, with the scenery unimpressive, there is little to watch except the four lanes with their constantly changing traffic, and the roadside succession of service-stations, "joints," and tourist courts.

The historical origins of this sector date back three centuries, to 1651.

At that time the international situation along the Delaware was complicated. Both English and Dutch claimed the region, but since 1638 the actual possession had been held by the Swedes, who had made several settlements—although, as an Irishman might say, "A lot of those Swedes were Finns!"

In the summer of 1651 the Dutch moved to assert their sovereignty. Their warships sailed up the river, and anchored off what they called Sand Hook. Soldiers and settlers went ashore, and began to construct a rectangular stockade that they named Fort Casimir.

With considerable probability of being right—and certainly with no risk of being proved wrong—one may assume the beginnings of U. S. 40 as of this time. The sand-spit upon which the fort was built has washed away, but it lay just off the present ferry-slips. The men who were building the fort had to bring wood from a little distance inland, and the line of Chestnut Street, by which motorists still (1950) leave the ferry, lies just where such a road would have run. The earlier names for this street indicate its antiquity and imply that it was the original east-west road. It was once called Thwart Street, apparently because it ran athwart the ordinary line of travel, that is, the river. It was also at one time called simply the "Street leading to ye woods."

From this time on, there was a small settlement, after 1656 called New Amstel, at the site of New Castle. These Dutch settlers presumably wandered in the woods occasionally, hunting. But there is nothing to indicate that they had any need to extend the road farther westward than the few hundred yards that it was probably opened at the time of the building of the fort.

The neighboring Swedes and Finns, however, were better woodsmen, and probably wandered more widely. Unlike the Dutch, and the English too, these Baltic peoples were forest-dwellers at home, and already frontiersmen, after a fashion, at the time of their coming to America. So, when a Swedish soldier became disgusted or when a Finnish colonist fell into debt, or if either a Swede or a Finn got into trouble with the authorities or even if he merely got itchy feet, he was likely to "take to the woods" literally, build himself a cabin, and live as a squatter and hunter. There is evidence that as early as 1653 some of these runaways had already headed west.

Two or three or half a dozen fugitives may "make tracks" in the figurative sense, but they hardly leave a permanent track that can be called a road. Not for several years did such running-away become common enough to be likely to beat down even a path, and thus extend the route of U. S. 40 some miles westward. Curiously, the extension came about from depression, not from prosperity.

In 1658 the settlements along the river, by now wholly under Dutch rule, began to experience hard times. That winter was very cold: "The Delaware was frozen over in one night, so that a deer could run over it, which the Indians relate had not happened in the memory of man."

The next winter was also a hard one, and caused suffering. In addition the crops were bad, and there were numerous deaths from an epidemic. Many people were oppressed by debt. The Swedes and Finns had no sentimental loyalty to the Dutch regime. So—suddenly, it would seem, in this year—the running-away took on wholesale proportions. The Dutch authorities became alarmed. Of fifty soldiers, a half had deserted. Along with them, into the woods, had gone debtors and others, women among them. This "elopement," as the older books call it, was the easier in that only a dozen miles westward the runaways could escape from Dutch jurisdiction by entering the English colony of Maryland. Under such unfortunate circumstances began the first migration westward along the line of U. S. 40.

The fugitives were fairly numerous; also, we can assume, they occasionally chanced it to sneak back, visit some of the outlying farms for supplies and a chat with some old friends who could be trusted not to turn them in to the officers. One would guess that some kind of trail was beaten down through the woods. If so, it can be considered the first extension of U. S. 40.

From this same summer of 1659 we have the first actual account written down by a man who went across from the Delaware to the Chesapeake. The keeper of the journal was not a fugitive; in fact he was an important person, no less a one than Augustine Herrman, sent as an envoy from the Governor of New Netherland to the Governor of Maryland.

Nevertheless, Herrman traveled in modest style. He left New Amstel about noon on September 30, and went west on foot through the woods with Resolved Waldron, his fellow-envoy, a small escort of soldiers, and some Indian guides. They followed compass courses. From the directions and distances noted in the brief journal, we should judge that they kept to the north of the present route of the highway. After traveling three hours and a half, they camped for the night by a "run of water flowing southwards."

Next day they went on, and reached Elk River in good time. As they had been expecting, they found a boat there, and though it was leaky, they managed to proceed from that point on, by water.

Like so many early journals Herrman's is tantalizing by its brevity. The finding of the boat suggests an established route of travel. On the other hand, Herrman nowhere mentions the existence of a path, even of an Indian trail, and for one section he definitely states that there was no path at all. Perhaps the fugitives had not been numerous enough to beat down a trail, or perhaps they had managed to keep their route a secret.

There can be no question about the fugitives, however, for on the next day Herrman met two of them. One of them was "Abraham the Finn." He had been a soldier, and had deserted, but Herrman calls him now "the hunter." The other fugitive was a Dutch woman whom Abraham had "brought hither"—so that this might be called an "elopement" pretty much in the modern sense of the word.

From the evidence of Herrman's journal, all in all, we cannot be certain as to whether there was even an established foot-trail from the Delaware to the Chesapeake as early as 1659. Probably there was, somewhere. But even if not, one must have been developed very soon after that time.

There began to be back-and-forth travel between the Dutch and English settlements. The runaway activity, for one thing, worked in both directions. There were indentured servants in Maryland, and these sometimes "eloped" eastward, seeking to escape English authority. Thus in 1660 Mr. Coursay, a gentleman of Maryland, came into New Amstel seeking three servants. Nothing is said about whether Mr. Coursay was on horseback or afoot, but knowing that he was a Maryland planter, we may think it likely that he was on horseback.

Actually, the route across from bay to river is so easy that there is no reason why Mr. Coursay should not have ridden horseback, even in 1660. Nothing in Herrman's journal, certainly, indicates any hardships of travel for a man on foot. The country is almost level, yet high enough so that there are no important swamps. In the great primeval forest the underbrush was probably not very thick.

From 1660 on, then, we shall not be stretching historical evidence very far to assume that a horsetrail led across between New Amstel and Maryland. Remembering that Herrman ended his land journey on Elk River, at or near the site of Elkton, we also have reason to assume that this route followed the general line of U. S. 40, westward from the Delaware, for fifteen miles.

The next stage, the development of a cartroad, is tied up—curiously enough—with this same Augustine Herrman. This Dutch envoy of 1659 was an interesting and capable fellow. He is generally called a Bohemian because he was born in Prague, but he was a German rather than a Czech. At least, he had no deep loyalty to the Dutch, and seeing much good land during his embassy to Maryland, he applied for a grant there, and in 1661 became the proprietor of a fine tract on the Eastern Shore. He called it Bohemia Manor, after his native country, and the estuary of Chesapeake Bay there became Bohemia River, as it still is.

Being energetic and full of business-sense, Herrman immediately began to plan a cartroad across to the Delaware. But we should look at the geography before we consider the temporary success and the ultimate failure of his road, for in its history is an excellent sermon upon the influence of geography, as opposed to human schemes, in the establishment of roads.

That geographical unit which has recently come to be known by the coined name Delmarva Peninsula includes approximately all of Delaware, together with the Eastern Shores of Maryland and Virginia. In shape it might be called bludgeon-like, being small at its base and bigger toward its end. From tidewater across to tidewater at the narrowest point is not over five or six miles. This narrowest portage lies between Appoquinimink Creek on the Delaware side and Bohemia River on the Chesapeake side. By making this short passage here you could avoid a long sea-voyage. One of the Dutch governors was so much impressed by the importance of this route that he considered abandoning New Amstel as his capital and building a new town on the Appoquinimink. The enterprising Herrman was also impressed, for there were obvious potentialities in the Chesapeake-Delaware trade.

Nothing came of the Dutch governor's project, but the possibilities for peaceful trade were increased by the English seizure of the Dutch settlements in 1664. Within a few years, Herrman had constructed a broad cart-road across the portage, and not long afterward this road was extended northward to New Castle, as the English had renamed the town.

According to the lights of his time, Herrman had built wisely. His road furnished the shortest possible overland link, and Maryland particularly was committed to the magnificent system of natural inland waterways supplied by the numerous branches of the Chesapeake.

But in the long run Herrman was wrong. The trouble with his route was that its western exit was wholly dependent upon water-transport. Now there are many occasions when travelers can't take a boat, or don't wish to. They may not have money to pay the passage, or they may have a herd of cattle or horses that are best transported on their own hooves, or the weather may be bad, or the boat busy or leaky. Or, as not infrequently happened even with very decent people in early times, the travelers may prefer to keep close to the shelter of the woods and not to expose themselves to anything so unpleasantly public as a boat-landing, where sheriffs can easily keep watch.

Here came the great and in the end overwhelming advantage of the New Castle and Elk River route, which is U. S. 40. It offered a two-way choice. If the west-bound traveler came to Elk River and wanted to take a boat, well and good! But if he preferred to go on by land, also well and good! For, if you look at any map, you will see that, keeping directly ahead at Elkton, you follow a road that neatly skirts the northern end of Chesapeake Bay, and all the West and South lies before you. This is just what U. S. 40 still does, and we can be pretty certain that for this reason an important road will continue to follow this route....

West of Elk River we can also sketch the development of the road. Possibly some sections here may actually have preceded the New Castle-Elkton section, for the records are so scanty as to make positive statements difficult. Any earlier development here, however, seems the more unlikely because of the Marylanders' devotion to travel by water. The colony did not even pass its first road law until 1666, and this is entitled "An act for making high wayes & making the heads of Rivers, Creeks, Branches and Swamps passable for horse and foote." The very wording implies that roads are considered a kind of secondary transportation-system, merely a linkage of the heads of navigation. Also there is no mention of wagons. The roads are merely to be made passable for men on foot and horseback.

In 1682 the Baltimore County Court began action against the overseers of highways in Gunpowder Hundred for not having their roads passable, but this complaint was dismissed later in the same year. By putting the two entries together we can assume that certainly by 1682, and probably earlier, there was some kind of road (for horses and for men, if not for wagons) near where the highway now crosses Gunpowder River, about twenty miles north and east of Baltimore.

After this time development was more rapid. By 1685 ferries had been established over Bush and Gunpowder rivers and a continuous road ran from the Patapsco, where Baltimore was later to be founded, to the Susquehanna. From that point a traveler should have been able without too great difficulty to get across to New Castle.

In 1690 the court of the colony had ordered that the roads should be made passable for carts. Such orders of early courts often express an ideal rather than enforce a reality. Nevertheless, the chances are that by that year you could have walked or ridden horseback and driven a pack-animal, and possibly have taken a cart, all the way from New Castle to the Patapsco, and probably have got some kind of ferry service across the larger rivers, including the Susquehanna. There were few, if any, bridges; across the smaller streams bridges would not be especially needed, and they were too expensive to build across the larger ones. The road itself must have been merely a forest-trail, from clearing to clearing, or—if you prefer to be a little grander—from plantation to plantation.

The whole road, in fact, was of so little importance that the first post-route, established in 1695, was taken across the bay from Annapolis to Chester River, and then continued north on the Eastern Shore, completely avoiding the head of the Bay.

In the early eighteenth century travel increased, and the New Castle-Patapsco route became a post-road. *Poor Richard's Almanac* of 1733 gives the itinerary along it, showing mileages somewhat greater than those of the present highway, as would be expected. In 1765 a stage-line began to operate, once a week, between Philadelphia and Baltimore, using the approximate line of U. S. 40 west of Elk River. In 1775 a stage-line was established between the same two cities, *via* New Castle and Frenchtown, which was then the village at the head of navigation on Elk River. Even yet, however, there was probably very little traffic and most of that on horseback.

Nevertheless, the road—primitive though it was, scarcely more than a track through the woods—was already the chief overland link of communication between the north and east of the colonies and the south and west. Little more than two weeks after the first stage-coach *via* New Castle, the rider bearing the news of Lexington, and war, must have gone squashing and splashing through the April mud, calling out the news at each clearing, spreading the word at the taverns and the ferries. Within the next year delegates to the Continental Congress from the southern colonies must have ridden northward along it, and passed many a rough-clad company of infantrymen plodding up from Virginia or even from the Carolinas.

In the next few years the little track through the woods saw other armies, and knew history being made. In 1777 Howe landed his British at Head of Elk, and marched along the road at least as far as Aikentown, since called Glasgow, where he set up his headquarters. In 1781 Lafayette marched south with an American force as far as Head of Elk, and there took boat for Annapolis. A year later the road felt the tread of Washington's Continentals and Rochambeau's French, marching south to a meeting with Cornwallis at Yorktown.

With peace and increased population and trade the road became more and more traveled, but did not improve its condition as a highway. In fact, since little work was done upon it, increased traffic probably made it worse, rather than better. Isaac Weld, one of those indefatigable British travelers, went over it in a public stage from Elkton to Baltimore in 1795, and in *Travels Through the States of North America* (1799) has left a description that has become classic for early American roads:

> The roads in this state [Maryland] are worse than in any one in Union; indeed so very bad are they, that on going from Elktown to the Susquehannah Ferry, the driver frequently had to call to the passengers in the stage, to lean out of the carriage first at one side, then at the other, to prevent it from oversetting in the deep ruts with which the road abounds: "Now, Gentlemen, to the right;" upon which the passengers all stretched their bodies half way out of the carriage to balance it on that side: "Now, Gentlemen, to the left," and so on.

Weld commented upon the inadequate methods of repairing the road, merely by piling saplings and bushes into the ruts and covering them with earth. He also noted that in the woods when one route became full of mud holes, it was customary to start a new track, so that six or seven roads might be seen branching out one from another but all returning eventually to the same route. By this time there were bridges over the creeks, but Weld describes them as merely covered with loose boards, and so weak that they tottered as a carriage passed over.

The *American Annual Register* for 1797 confirms, or underlines, Weld's description: "[Roads] from Philadelphia to Baltimore exhibit . . . an aspect of savage desolation. Chasms of the depth of six, eight, or ten feet occur at numerous intervals. . . . Coaches are overturned, passengers killed, and horses destroyed by the overwork put upon them."

In the first decade after 1800 the steamboat suddenly brought water-transport back into favor. The best route for travelers came to be—water from Philadelphia to New Castle, road across to Frenchtown, water again to Baltimore. As a result, west of Elk River the road was less traveled than before, but between New Castle and Elk River it had a short season of prosperity.

A turnpike company, to construct an improved road from the Delaware to the Chesapeake, was chartered in 1809, and the turnpike, sixteen and one half miles long, was finished by 1818. Thus, more than a century and a half after Augustine Herrman journeyed across the narrow isthmus, the road was again serving principally as a link between two systems of water-transportation. But the end was at hand.

In 1827 the turnpike company got its first authority to build a railroad along its right of way, and in 1830 it became the New Castle and Frenchtown Railroad Company. In the next year a very primitive railroad, using horses, started operation.

In 1837 steam trains began to run from Philadelphia, *via* Wilmington and Elkton, to Baltimore. At that time the through travel over the stage-road must practically have ceased.

During the Dark Ages the competition of two paralleling railroads and of water-transportation on the Bay sent this sector of the road well back into the mud. Only small towns and a not very prosperous farming country, reverting to woodland, lay along the course of the old road, and so there was little local travel. A detailed map of Maryland in 1899 shows good turnpikes radiating far out from Baltimore toward the west, northwest, and north, and one of them even following the line of U. S. 1 northeastward for twenty-five miles. But along the route of U. S. 40 there is only the broken line of an abandoned toll-road as far as Gunpowder River, and beyond that no continuous road at all.

Out of this very degradation, curiously, has sprung some of the present prosperity of U. S. 40. When the numbered highways were established, the proud designation U. S. 1 was given to a road that was intended to be the chief route of north-south travel through the great cities along the Atlantic seaboard. And so it is, over most of its length. Between Philadelphia and Baltimore, however, U. S. 1 was properly assigned to what was the best road at the time. Its general route, however, proved to be inferior to that of the old road around the head of the Bay. As a result, U. S. 40 has been transformed into a modern super-highway, of four lanes, sometimes of six, and U. S. 1 is left, in this sector, a secondary route. Moreover, U. S. 40 is here transformed into a north-south rather than an east-west highway. By means of its connection with U. S. 13 it is the chief road from Philadelphia to Baltimore, Washington, and the South. Even traffic from New York, keeping to the Jersey side of the Delaware, by-passes Philadelphia, and uses U. S. 40 to reach Baltimore.

This southward dip to Baltimore is, as far as the transcontinental aspect is concerned, an anomaly that results from past history, and is eloquent of the manner in which the system of numbered highways was constructed out of old routes, with merely a city-to-city ideal. The head of the Chesapeake is almost directly east of the line of U. S. 40 at Hagerstown. There is no important natural obstacle, but no modern highway goes that way, presumably because no old line of travel developed in that direction. On the other hand, a road directed straight west from Baltimore would strike U. S. 40 at Vandalia, thus avoiding the northern detour that it actually takes. Just how much money has been expended in extra mileage of highway construction and how much money is consumed yearly in gasoline and tires because of such wanderings of our highways is a statistical question involving too many factors to be clearly answered. Our modern routes are actually the result of a compromise between the always inescapable facts of geography and the equally inescapable hand of the past.

As the result of straightenings in the last twenty years the new modern highway has been largely relocated, and the old road of colonial times has now become Maryland 7. Yet, though relocated, the new highway still follows essentially the old route, and is rarely much more than a mile distant from it. Geography still determines. U. S. 40 remains, as it here began, a road which from the Delaware-crossing at New Castle skirts the northern end of Chesapeake Bay, and keeping just at the heads of the estuaries runs south and west. . . .

Change continues. The opening of the Delaware Memorial Bridge and its approaches marks the latest relocation. By crossing the river about two miles north of New Castle, the highway leaves the line of the "Street leading to ye woods," but the spires of the town remain in sight.

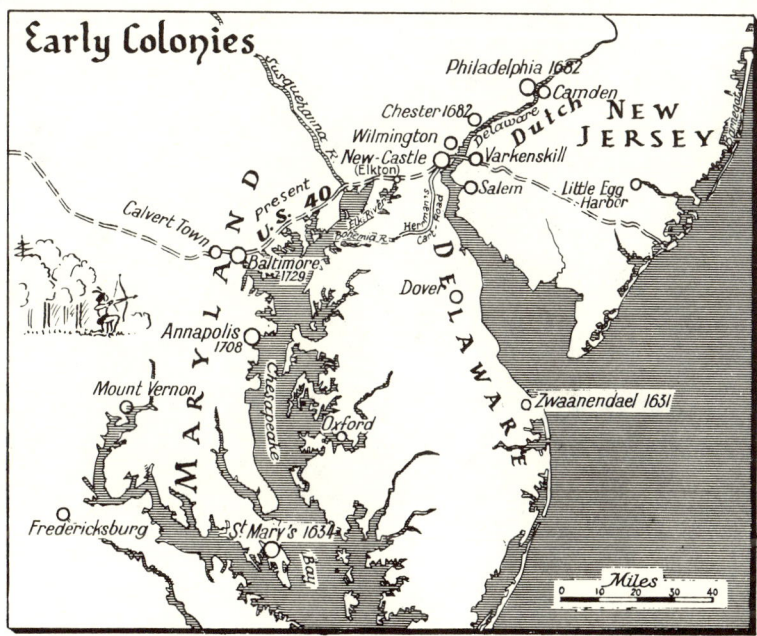

# 4 New Castle

**(40)** In the summer of 1950, although the building of the Delaware Memorial Bridge two miles northward was progressing rapidly, U.S. 40 still was broken in its continuity—the only break in its whole length—by the water of the Delaware River between Pennsville, New Jersey, and New Castle, Delaware. For sentimental reasons we may be glad that the bridge was not as yet completed, for at New Castle U. S. 40 reaches back deeply into the American past, and from the ferry the traveler has a glimpse of the highway's very beginning in time.

On this particular afternoon, when the west-bound ferry-boat was approaching the slips, it reversed propellers to allow another boat to get out of the way. The quiet and muddy waters of the river are thus thrown into a sudden turmoil marked by patterns of foam at a historic spot.

Just about at this point the Dutch ships probably rode at anchor in the summer of 1651. Immediately ahead, where now is only water, lay the spit of land called Sand Hook, on which the Dutch built their Fort Casimir. From beyond the ferry-slips, angling a little to the right, now runs U. S. 40, and there once presumably ran the "street leading to ye woods," which in turn, we may believe, developed from the track laid out when the Fort was being built in 1651.

From Fort Casimir also, on that September 30, about noon, in 1659, Augustine Herrman set out westward on the first recorded journey along the line of U. S. 40.

Although the original woods have all long since been cut, the number of individual large trees still showing their rounded tops along the horizon line is notable, and gives some indication of what the first voyagers must have seen.

The town itself shows but slightly in the picture, but New Castle, with buildings dating from the mid-seventeenth century, is actually one of the architectural gems of the country. One or two of the fine old buildings along the town's famous Strand may be seen between the ferry boats. The graceful steeple is that of Immanuel Church. The establishment of this parish dates from 1689, part of the present building from 1703, and the steeple itself from 1820-22.

One of the few details of this picture that would have been familiar to the early voyagers is the accommodating seagull that flew across just in time to relieve an otherwise grey sky.

# ⑤ Six-Lane Highway

🛣️40 Two and a half miles west from the New Castle ferry-slips U.S. 40 joins U. S. 13, and the two highways coincide for about a mile. Except for strictly urban development this is the only part of U.S. 40 that at present attains six-lane width, and it is the section that carries the heaviest traffic. According to the latest official count the average day's traffic at this point totaled 22,688 vehicles. This is borne out by the actual picture, in which ten cars and two trucks appear, even though only a short stretch of highway is visible. Almost half of this traffic is apparently to be credited to U. S. 13, since at the separation of these highways, only a mile to the southwest, the traffic following U. S. 40 is reduced to 12,853.

This short stretch of U.S. 40 here makes use of the Coleman du Pont Road, commonly called the Du Pont Boulevard, constructed between 1911 and 1924, largely at the initiative of, and from contributions by, Mr. Coleman du Pont. Like the initiators of the Lincoln Highway, which slightly antedates his project, Mr. du Pont was attempting to establish "the road of the future," and he was also desirous of benefiting the people of his native state.

Original plans called for a right-of-way two hundred feet wide, an almost astronomical figure for those days but by no means excessive according to present standards. Unfortunately this was later somewhat reduced, and the right-of-way as shown in the picture is somewhat less than two hundred feet. It consists actually of three northbound lanes totaling thirty-six feet, three southbound lanes totaling thirty-eight feet, and a median strip seventy feet wide, plus rather inadequate shoulders.

In spite of its six lanes and wide dividing strip, The Highway of the Future has in less than forty years become a highway of the past by failing to attain the highest standard of modern design. This is largely because it was originally planned as a four-lane road, and the outside lane, when added, was cramped and had to encroach upon the shoulder. As a result the curb and mailbox at the right come much too close to the line of the highway, interfere with traffic in the outer lane, and might well cause accidents.

The highway here partakes of the parkway type of road, with the grass and tree-plantings of the median strip carefully maintained. It displays therefore something of the generally artificial appearance of such parkways, and is "dominant" in forcing itself upon the countryside rather than mingling with it.

Actually, the countryside here is merely the gently rolling coastal plain, two miles inland from the Delaware estuary. As the background shows, it is a good tree-growing country, and in a generation or two the now small trees growing in the median strip will have attained mature magnificence.

## ❻ Bush River

🛣️40 About twenty-five miles east of Baltimore, U. S. 40 skirts the head of the estuary formed by one branch of Bush River, an arm of Chesapeake Bay.

The picture was taken from a point about twenty feet above the highway on the embankment of the railway that here parallels it. The direction of view is about southeast.

The highway itself is here a good example of four-lane construction. The pavement is of concrete, with the asphalt-filled joints showing clearly.

The separation-strip is of adequate width, and the grass is well cut. A drainage sump is provided. The shoulder is of good width, partly surfaced with bituminous material and partly graveled, providing plenty of room for cars to pull away from the pavement.

The topography is that of the so-called "drowned" coastal plain. Perhaps not more than twenty thousand years ago the area now represented by the estuary was a nearly level valley with a small stream meandering down through it to join the larger river where the channel of Chesapeake Bay now lies. When the melting of the great ice-cap released tremendous amounts of water, the sea flooded back along these valleys, transforming them into shallow bays. Since the original flooding the growth of marshy plants and a little filling in of sediment has probably enabled the land to advance again somewhat upon the water. At one time the tide may have reached up as far as the highway.

At the time of the coming of the white men heavy forest probably grew down toward the water's edge at least as far as where the line of high bushes now stands. The older trees have now been completely stripped off, and a vigorous growth of young elms is taking over. The elm, which has a wind-borne and easily sprouting seed, often manages to become dominant immediately after such trees as oaks and pines have been cut out.

Historically, U. S. 40 at this point makes its earliest contact with recorded history. On July 28, 1608, Captain John Smith set out on his second voyage of exploration on Chesapeake Bay. A few days later he entered an estuary which he called "Willowbyes river," and which can be positively identified with the present Bush River. Whether he ascended this branch or the more southerly one is not certain, yet it is at least no wholly unpardonable flight of fancy to imagine that his little shallow-draught "barge" once came feeling its way up the shallow channel beyond the farther trees.

Thirteen men were packed into her, but eight of them were down and out with a sickness, probably dysentery. In Bush River the Englishmen encountered several canoes full of Massawomeck Indians, who seemed to be advancing with hostile intent. Smith propped up the sick men's hats on sticks along the gunwale, and the Indians drew off and contented themselves with staring amazedly at the spectacle of a sailing vessel. Smith steered boldly down upon them, and by this firm front won their respect and friendship.

# 7 Baltimore Rows

In Baltimore, U.S. Alternate 40 runs between characteristic "rows." The view is east, toward the tall buildings in the center of the city. The day is overcast and drizzly; the streets, shiny with wetness.

Traffic is here controlled by lights, and in the immediate foreground the street is vacant. A double front of cars, however, can be seen charging half way down the block, having just been released by the change of light.

Waterloo Row, built in 1819, shortly after the battle from which the name was derived, introduced this style of architecture in Baltimore. The architect was Robert Mills (1781-1855), who thus left his mark, upon the city and upon the whole eastern seaboard.

The houses in the picture, with their even height of three and a half stories, brick construction, narrow fronts, and marble steps, are typical examples of rows. Baltimoreans take them for granted. Visitors are amused or horrified.

Certainly there are too many of them and they become monotonous. But actually much can be said for the warm reds of their honest brickwork, for their simple doorways—sometimes square-cut and corniced, sometimes arched,—and for the generally fine proportions of their facades. Moreover, they represent a real architectural tradition, developing out of a way of life that endured with much stability and homogeneity over several generations. We may contrast the more recently built districts of many cities where a dozen rootless imported styles of imported architecture clash in the same block.

In addition, for its time the row was highly practical, even functional. The height is suitable for buildings without elevators. Brick cut fire-hazard, and utilized good local clays. Narrow fronts kept distances from being too great in the times before rapid transportation.

Traditionally scrubbed every morning, the marble steps wear rapidly, and are also eaten away by the carbon-dioxide-laden air of the city. Definite sags show in the surface of the first stairway on the right, and the four upper steps of the second stairway have apparently replaced original ones, and contrast with the lowest step.

Beyond the fourth stairway a Gothic churchfront also abuts on the pavement. Even this church seems to adjust itself to the spirit of the row by presenting only a narrow front, and by seeming to reproduce the lines of the steps with its buttresses.

The littered street and the garbage-can at the left show this block to be possibly on the downward path, but the sprucely dressed little boy—obviously posing, but for the moment forgetful of his toy pistol—indicates that prosperity has not yet altogether moved around the corner.

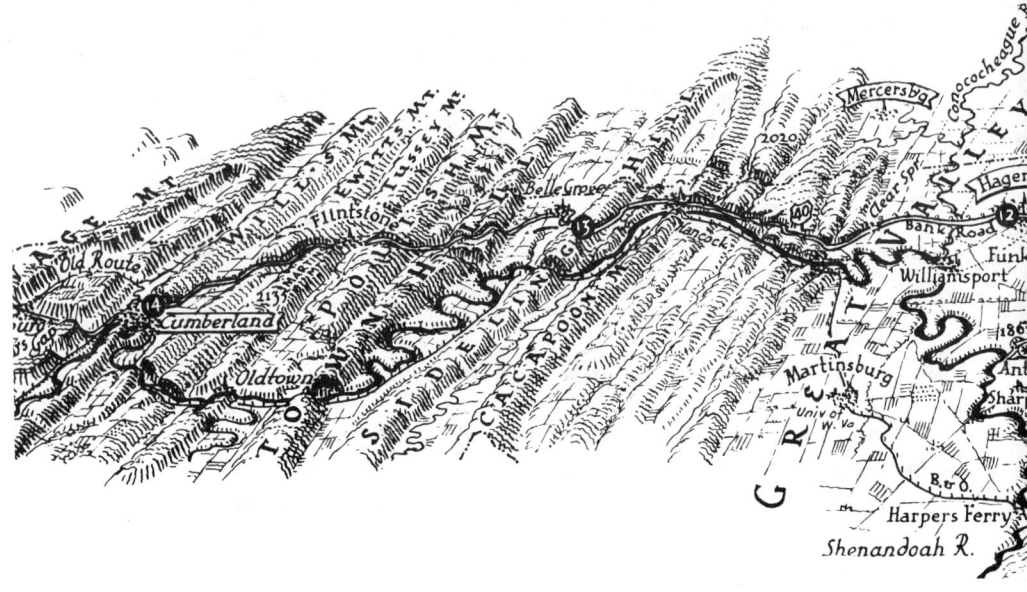

# BANK ROAD AND TURNPIKE
## Baltimore to Cumberland

The highway, turning westward at Baltimore, leaves the Coastal Plain. Its next sector, determined partly on topographical but more on historical grounds, extends to Cumberland in western Maryland, 134 miles.

Throughout this whole distance U.S. 40 traverses rolling and hilly country. As far as Frederick it passes through the low hills of the Piedmont. Next

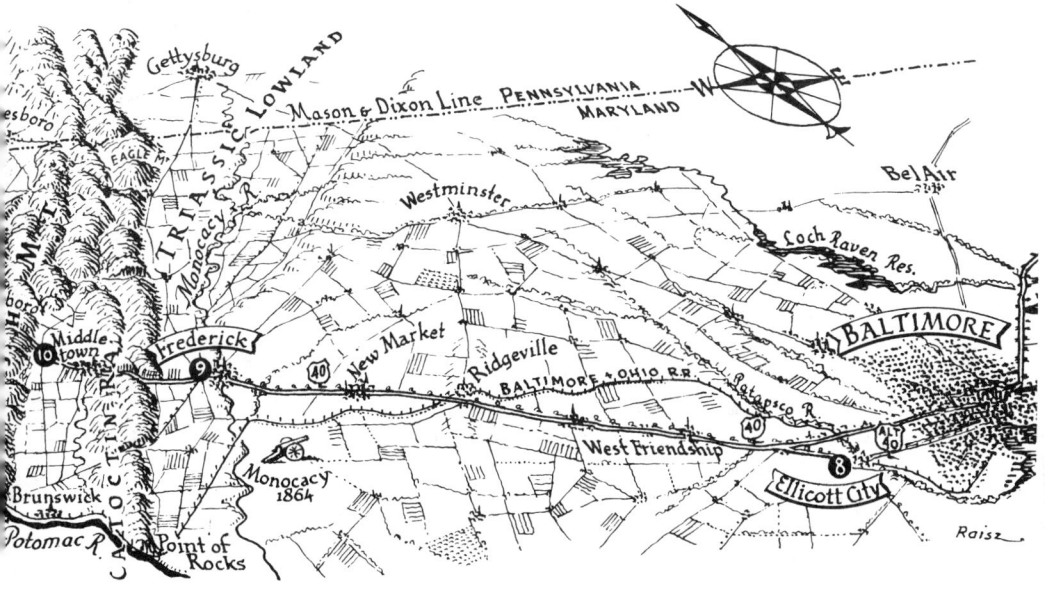

it crosses the Blue Ridge Mountains, here represented by the parallel ridges of Catoctin and South mountains. Still farther west it traverses the full breadth of the so-called Ridge-and-Valley Province, alternately ascending the high north-and-south-running ridges and dropping into the lower but often hilly valley that lies between.

Scenically, the run is varied, nearly always beautiful, and sometimes magnificent. In the broader valleys the farmlands are rich. The ridges are luxuriantly green with a thick forest of oaks and other broad-leaf trees.

Historically, the sector represents a kind of weak link—between tidewater at Baltimore and the beginning of what was first Braddock's Road and later the National Road, at Cumberland. This part of the road is, curiously, a little younger than the part farther west. It is also about a century younger than the part farther east—another reminder of how slowly the frontier advanced westward in colonial times.

By the middle of the eighteenth century various sections of road existed to the west of Baltimore, but they were of local significance, and had not been linked together to form a through route. This is not surprising. Baltimore, not even founded until 1729, failed to exercise much influence until after the middle of the century. Maryland was a small colony, and did not press westward as did Pennsylvania and Virginia. Even geography worked against Maryland, for there was easier access to its western lands from either of the neighboring colonies than across Maryland itself. Thus, the first road into Frederick came from the northeast and connected with Philadelphia, and the first road into Cumberland came from Winchester in Virginia.

Braddock's campaign, in 1755, emphasized the lack of communications in Maryland. At that time there was some kind of road between Baltimore and Frederick, and the general with a detachment of the army marched through Frederick. He did not, however, come from Baltimore, but from the south. Moreover, west of Frederick, his army had to cross the Potomac, march through Virginia, and then cross back near Cumberland.

Braddock met defeat, but his road remained as a valuable asset to Maryland. In 1758, after the war had taken a turn for the better, the colony set out to establish a road to connect with Braddock's. By this time a road had been pushed a dozen miles west of Hagerstown, and some work had been done east of Cumberland also. A gap of about forty miles had to be closed. The Assembly voted funds, partly for military reasons, partly to "induce many people to travel and carry on a trade in and through the Province, to and from the back country."

Presumably some kind of road was constructed at this time, but it must have remained in a primitive state for many years. Most of its traffic would have consisted of pack-horse trains, and Philadelphia continued to absorb the bulk of the western trade. The road from Hagerstown to Hancock was one of several that in 1790 were declared "much out of repair."

In 1792 the Legislature saw fit to pass a bill establishing as a public highway that road which "from time immemorial" had led between Baltimore and Frederick. The act is of interest for three reasons. First, it shows that the road, up to this time, had "just growed," without any official status at all. Second, the need of passing such an act probably shows that some private interests were threatening to control the road, doubtless by setting up tollgates along it. Third, the act shows that "time immemorial" according to American standards of the eighteenth century need not be longer than fifty years, for this particular road could not have been older than that!

Certainly an interest in toll-roads was already in the air. A scheme for a Baltimore-Frederick turnpike dates from as early as 1787, but nothing came of it until much later.

The authorization of the National Road in 1806 dropped a political plum into Maryland's lap, but to gain the full advantage for the now booming city of Baltimore, the connecting road must be improved. One method adopted was the establishment of turnpikes, and the first twenty miles of such a road, west of Baltimore, were opened to traffic in 1807.

The construction lagged, however, and in 1812 the state adopted an unusual and ingenious plan. In connection with reincorporation of various state banks, it was required that they finance and construct the road. This device seems to have worked fairly well, and a marker on U. S. 40 west of Hagerstown still declares this to be "The Bank Road," and adds "The portion of this highway from the west end of the Conococheague Bridge to Cumberland (40 miles) was built between 1816 and 1821. The banks of Maryland financed it by purchase of the stock."

Thus completed, the Baltimore-Cumberland sector merges historically with the National Road. Together they connected the West and Baltimore, and largely because of this western trade Baltimore for a while surpassed Philadelphia and was second only to New York among our cities. "In 1827," the *Niles Register* records, "a gentleman traveling thirty-five miles on the road between Baltimore and Frederick met or passed 235 wagons . . . nearly seven for every mile. These wagons were generally of the largest size and very heavily loaded."

Even at this peak of greatness, however, the future of the road was threatened. In this same year the Baltimore & Ohio Railroad was incorporated. The railhead reached Ellicott City in 1830, and was then pushed westward, transforming the road to a secondary means of communication. The trains began to run to Cumberland in 1842.

By the building of the railroad Baltimore maintained its touch with the West. Even during the railroad era, however, the fine turnpikes serving the city did not fall wholly into disrepair, and the Dark Ages were not as dark in Maryland as in most states.

An official report of 1899, for Howard County, just west of Baltimore, is of interest as showing the general conditions of that time. Teaming had originally been chiefly east and west along the line of the turnpike. It was noted, however, as now being chiefly north and south, and as consisting mostly of short hauls, to and from the nearest railroad station. Nevertheless much of this hauling had to pass along the turnpike for short distances, and the tolls were still enough to maintain the roadway in good condition.

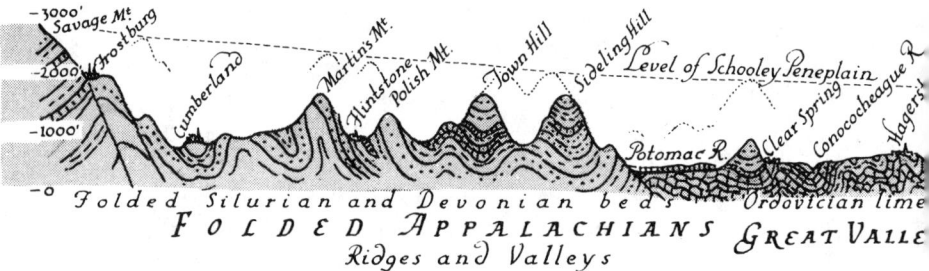

*Baltimore to Cumberland · 71*

At the end of the century toll-roads still extended all the way from Baltimore to a point seven miles west of Hagerstown, a distance of about eighty miles. Some of them continued to function well into the nineteen-hundreds.

From Baltimore to Hagerstown the highway has been largely rebuilt for modern traffic, and has been relocated. Its unrelocated sections, however, still follow the lines of the turnpikes. West of Hagerstown its route is still essentially that of the Bank Road.

The location of this sector seems to be the result of political and commercial pressures more than of geography. It runs largely, so to speak, across the grain of the country, and a better route might well have been selected. What was wanted at the time, however, was an all-Maryland road, to bring the commerce of the West to the wharves of Baltimore, and this is what U. S. 40 has inherited.

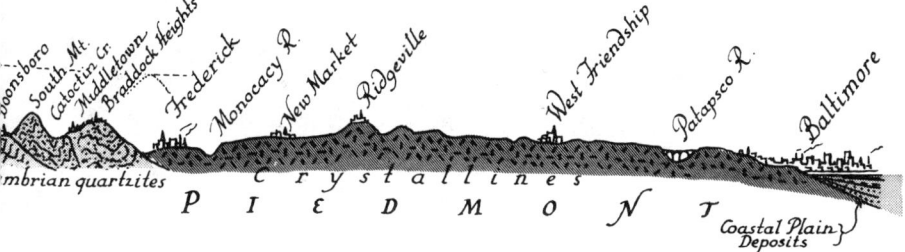

## ⑧ Ellicott City

🛡40 The narrow, winding, heavy-graded, main street of an eighteenth-century Maryland town, along which U. S. 40 still passes, shows why the life of a highway commissioner in these United States is seldom an easy one.

Like most towns in the southern colonies, Ellicott City merely "grew." There is no village green, no planned layout of streets, such as one finds in

New England. In the South the plantation came before the town, and the original road took the easiest course from plantation to plantation.

In the beginning, obviously, the town had no need for a wide street, and the right-of-way from building to building was established at about forty-five feet. Three feet of this on each side were allowed, by custom, to be used for projecting steps, as in the Baltimore rows, and for shop-windows, set forward from the building-front, as in old English towns. An additional four feet on each side, were reserved for sidewalk. When space is then left for parking, the effective width of the street is only about twenty feet, for street-car tracks and all traffic.

In the end, as frequently happens, the very badness of the situation has brought about a change for the better. Modern highways have had to be relocated entirely, so as to pass outside the towns. This is what has happened with Ellicott City, and the through traffic has been diverted over Alternate 40, which passes a mile or so north of the congested main street here shown.

The center around which Ellicott City developed was the grist mill, founded in 1774 by the three Ellicott brothers. They located their mill at a point where it could utilize the water power of the Patapsco River, which here tumbles down across some obstructions of granitic rock. This rock itself appears in the façades of the two massively built old buildings at the right of the picture. The height of these buildings, which seems excessive for a small town, is partly the result of the location in the ravine, along the stream. The closer building, the masonry of which has recently been re-pointed, is constructed of irregularly shaped stones. The farther one is of immense rectangular blocks. This building also shows, except in its top story, the old-fashioned small window-panes, a legacy of the period when glass was expensive and when glass-makers had not mastered the art of making large panes.

The excessive cluttering of streets with poles and wires, so common in American towns, shows well in the picture. On the right, tall poles carry a three-wire electric power-line with its rather heavy insulators showing against the sky. Wires for local distribution of power, with smaller insulators, are carried on the lower cross-arm. To the left, a single cable of telephone-wires is carried on short poles. Wires for the trolley-car further complicate the pattern.

# ❾ Frederick

🛣️ 40 The winding street of historic Frederick—like that of Ellicott City, following the narrow right-of-way of an old road—is interesting enough in itself, and the Barbara Fritchie Museum offers a point of focus.

Actually, one has a little difficulty in saying just what the museum represents. It is not the original house, being—as the W.P.A. Guide circumspectly declares—"supposedly a reproduction." Moreover, one has some difficulty in deciding exactly what is here commemorated, except some verses by a New England poet. Certainly it seems extremely doubtful that any woman, "bowed with her fourscore years and ten," defiantly hung out her stars-and-stripes from a window located where the flag now hangs, and that Stonewall Jackson threatened his men:

> "Who touches a hair of yon gray head
> Dies like a dog! March on!" he said.

It seems also unlikely that the Confederates marched "All day long through Frederick street" beneath that hostile banner, and even Whittier ignores the practical difficulty, that is, the staff being shattered, did Barbara stand there and hold the flag all day, or did someone bring her up a new flagstaff?

Still, it is a fine American legend. No less a personage than Winston Churchill showed himself familiar with the poem by quoting it freely, on the occasion of his visit to the museum. Even if things did not happen just as related, the poem has become a part of the American background, and thousands of people pass through the rooms of the little cottage here pictured, where they see a few relics of Barbara, and see also some relics of the Civil War battles of South Mountain and Monocacy, both fought in the vicinity.

Moreover, there is no doubt that this street has been a thoroughfare of war. One of Braddock's red-coated regiments marched westward through Frederick in 1755. Blue and gray columns marched and countermarched along it, and Stonewall Jackson must have ridden beneath Barbara's window, even if he did not stop, first to order a volley at the flag, and then to have a change of heart.

Unpretentiously genuine is the old painted-brick house beyond the museum. Dating probably from the early nineteenth century, it is a typical town-house of provincial Georgian, broader-fronted than the Baltimore houses, because built in a smaller town where space and distance were not so important. Nevertheless, it was built with the idea that it would stand in a row of houses, and so it has no end-windows, except for one squeezed into the garret. Such a house was certainly built without an architect, and its good taste and fine proportions spring from the long tradition of English Georgian building. Since the width of the front allowed room for only four windows upstairs, the builder made a virtue of necessity by placing the door in the position of the second window, thus achieving an interesting off-center effect. The second house beyond is also old, although its arched windows and doorways would indicate a slightly later date.

The television aerials supply the most modern touch. These were erected between the time when I took a picture in 1949, and the present picture, of 1950. The Francis Scott Key Hotel, boldly proclaiming itself on the skyline, commemorates the birth in Frederick of the author of the national anthem.

## ⑩ Maryland Countryside

🛣️40 "Fair as the garden of the Lord," wrote Whittier of the country around Frederick, and anyone standing on the slope of Catoctin Mountain and looking out toward South Mountain is inclined to agree with him.

Topographically, this view represents the beginning of the Appalachian Highlands. Catoctin Mountain is the first outlying bastion; the solid wall of South Mountain represents the beginning of the real Blue Ridge area. Both mountains are composed of harder rock, and the softer rocks between them have been worn away to form the fertile valley.

On the slope of the hill in the foreground, corn has been planted in strips following the contours, to reduce erosion. This contour-planting shows again, even more prominently, with the corn on the slope to the left in the middle distance. Already this method of plowing, which has so greatly changed the whole pattern of many parts of the American countryside, is coming to seem a chief legacy of the New Deal, probably to be vividly in existence centuries after NRA and AAA are dim initials in the history books.

Across the valley, pasture mingles with plowland. Since it is September, the wheat has been cut, but a field of oats shows light-colored near the middle of the picture. Well beyond, the village of Middletown is a cluster of white spires and buildings among its dark shade trees.

Pasturelands run well up the easily sloping sides of South Mountain, here and there actually touching the ridge.

Since Maryland lies east of the township-section surveying area, the fields are irregularly shaped. A general hedgerow-effect emphasizes the irregular field pattern, suggesting the rapidity with which this whole eastern part of the United States would return to forest if the hand of man were removed. These trees have not been planted, but have merely sprung up along fences where the sprouts were protected from the plow.

Many isolated farms show in the picture, as is to be expected in the United States, where, in spite of the frequent Indian menace, the European system of living in a small village and working out from that point never managed to establish itself.

The sky, as if attempting to match the landscape in delicacy, shows an overcast of strato-cumulus, thinner in some spots than others, producing an opalescent tone.

The picture was taken from the edge of the highway, which here swings to the right down the hill and then re-appears crossing the valley, rising and falling with the natural roll of the valley land. It here follows one of the old turnpikes. The newer, high-speed highway, Alternate 40, runs along a more direct route several miles to the north.

Along the road in view, on September 13, 1862, Union cavalry attacked Confederate cavalry in a series of engagements. The latter, fighting a delaying action, were gradually driven back through the village, and by evening had retired to the edge of the valley. In the hard-fought battle of South Mountain, the next day, the Army of the Potomac forced the gap of the mountain along the line of the road, and the die was cast for the great battle of Antietam a few days later.

General Jesse Lee Reno, commanding the IX Corps of the Union army, was killed at this battle, which straddled U. S. 40. By coincidence, his name stands on the city of Reno, Nevada, near the other end of U. S. 40.

## ⓫ Horrible Example

🛣️40 The westward-traveling motorist, turning sharply to the right as he leaves the village of Funkstown, then swerving again, crosses the old bridge over sluggish Antietam Creek. Both road and bridge are historic reminders of old Maryland.

The road in this section was constructed as a turnpike in 1822, and was the last gap to be finished in the linkage of the National Road with Baltimore. It was thus one of the chief connections between East and West until the completion of the railroad. During the Confederate invasions of the North, in 1862 and 1863, both the armies used it as a thoroughfare. The bloody two-day battle of Antietam, fought largely for the possession of bridges not unlike this one, occurred a few miles to the south.

The bridge itself was built in 1823, being the first of a large number of fine stone bridges to be constructed across the Antietam. The region supplies good limestone, and its early settlers were largely German in origin, with a tradition of solid masonry behind them. This particular bridge has been resurfaced, widened, and given a new parapet, to accommodate modern traffic. Nevertheless it still rests solidly upon the original three stone arches. Civil War skirmishes were fought at this point, and attempts were made to destroy the bridge. If it had been of wood, it would have been undoubtedly burned. Solidly built masonry, however, could resist the ordinary field-artillery of the time, and only a siege-train or some concentrated effort of the engineers could have destroyed it. . . .

The billboards ruin everything. The historical flavor, the old-time architecture, even the beauty of the wooded hillside—all are sacrificed.

Pole-lines and wires may be accepted, like fences, as part of the basic American landscape. They do their work without striving to be conspicuous, and often their not-ungraceful curves add a touch of interest, an intricacy of pattern, even some beauty. Billboards are different. In this picture, for instance, a three-wire power-line and two telephone-cables are inconspicuous except perhaps when the power-line insulators break the sky-line. But the billboards blast themselves into the viewer's consciousness.

During several trips across the country the photographer has studied billboards, and has tried to assimilate them, as he has done with the pole-lines, to the whole idea of the cross section. He must admit, in general, failure. The best that he can say is that some of the smaller billboards—those advertising local hotels, service-stations, or small industries—seem to have a certain rooting in the soil, and are often modest and comparatively harmonious to the setting.

The large billboards—owned by special companies, usually advertising the products of mass-production—are always placed in the most conspicuous spots, and have designs and colors carefully chosen to clash with the background. One feels a difference between a home-produced: "Stop at Joe's Service Station for Gas—Two Miles," or "The Liberty Café —Short Orders at All Hours—Give Us a Try!" and some gigantic rectangle advertising tires or beer.

Large billboards are now springing up along U. S. 40 even in the vastnesses of the Nevada sagebrush country. They are an abomination! Personally, I try to buy as little as possible of anything that is so advertised.

## ⑫ Mount Prospect

🛣️ Mount Prospect, 201 West Washington Street, in Hagerstown, Maryland, was built in 1789 by Nathaniel Rochester, who had served in the Revolution as a paymaster. His name is widely known because at a later date he founded the city of Rochester, New York. His Hagerstown residence proves Colonel Rochester to have been a man of taste, and it is as handsome a historic structure as anywhere faces the three-thousand-mile course of U. S. 40.

Its date places the house in the earliest years of the Classical Revival, and its portico shows that influence. On the other hand, its slightly arched doorway and first-story windows indicate a French influence, which was also strong in the years succeeding the French alliance during the Revolution.

Mount Prospect, built on a slight rise, must originally have deserved its name. Now, however, the view has been largely cut off by the growth of the town around it. The stone stairway, brick sidewalk, and stone curb are in the early American tradition. The iron fence and gate with its elaborately designed gateposts are Victorian touches. The parking-meter is wholly modern in function, but shows less modernism in its design than might have been expected.

The house is the scene of an incident involving a famous American son and described by the pen of a famous American father. At the battle of Antietam in 1862, fought a few miles away, a young Massachusetts captain was slightly wounded in the neck. During the Civil War surgical techniques were still primitive, and discipline for a wounded officer was easily relaxed. The captain made his own way to Hagerstown, intending to take the train from there to Philadelphia, where he had friends.

Let us now allow the father to take over—

> But as he walked languidly along, some ladies saw him across the street [perhaps from where the parking-meter now stands], and seeing, were moved with pity, and pitying, spoke such soft words that he was tempted to accept their invitation and rest awhile beneath their hospitable roof. The mansion was old, as the dwellings of gentlefolks should be; the ladies were some of them young, and all were full of kindness; there were gentle cares, and unasked luxuries, and pleasant talk, and music-sprinklings from the piano, with a sweet voice to keep them company.

Obviously, there should have been love and a marriage, but real life does not always follow the proper pattern of romance, even when it seems about to do so. In any case, the young captain survived his wound, survived also three more years of war, and lived to such an extreme and sprightly old age, still keeping his youthful interests, that at some time he made the famous epigram after passing a pretty girl: "Oh, to be seventy again!" Need he be identified as Oliver Wendell Holmes, Jr.?

# ⓭ Ridge and Valley

🛣️40 The view is eastward, from where U. S. 40 crosses the summit of Town Mountain, toward Sideling Hill.

The long and level skyline, formed by a narrow ridge, is typical of the Ridge and Valley Province, and indeed of the whole Appalachian region. Although the details of the geological process producing such a landscape are difficult and even disputed, the general outlines are clear. There was once, geologists accept, an almost level plain over this area. This plain existed at a rather ancient era, and was almost at sea-level. It was then raised, without much warping, until the whole area was, theoretically, a plain at about the present height of the ridges. Vigorous erosion, by means of streams, then began. The plain consisted of narrow bands of hard rock at the surface, separated by broad belts of soft rock. Gradually the soft rocks wore away, and the hard rocks were left standing as the tops of ridges, their height still remaining about as it had been at the beginning. Thus both Town Mountain and Sideling Hill are about 1600 feet above sea-level, and are capped with long strips of the hard Pocono sandstone of the Carboniferous period, while the valley between them has been worn down several hundred feet through the soft Devonian formations.

These ancient plains, now showing only as the level tops of ridges, are so well marked that they have even been named. The one here showing is known as the Schooley Peneplain, and can be traced over much of the Appalachian area.

The slopes of the ridges are generally left in forest. Oaks are probably the most common tree, but many varieties occur. There is a fair amount of sugar maple, and much sugar is produced along U. S. 40, some of it to be sold later under Vermont labels. Chestnut was common until the blight exterminated it.

The valleys are often partly under cultivation, as here. The number of farms and barns visible in the sunlight-spotted landscape would indicate a fairly rich soil.

Sideling Hill, in its very name, displays the nature of its relationship to the road. It could not, like many ridges, be passed by a gap, or ascended along the channel of a stream, or up a minor ridge. It was also too steep to be easily climbed directly, and the name—recorded from before 1760—precedes the opening of a wagon-road, and probably commemorates the route of the first horse-trail. Packers must have had to work up the slope by "sideling" along it, and hence the name. Even on the modern map the position of Sideling Hill is marked by a very distinct V in the line of the highway.

Even when relocated U. S. 40 will ascend the ridge by a route cut out along the side. In fact, the sharp white scar on the slope shows the line of the new highway, which runs parallel to and just above the old Bank Road, the present highway. In a country where growth is so rapid, such a scar will not long remain very prominent. The line ascending the ridge vertically, just to the right of the highway, is the route of a powerline.

The picture, with its variegated pattern of light and shade, was taken during a break in a day of heavy rain. The threatening clouds are apparently not more than a few hundred feet above the top of the ridge.

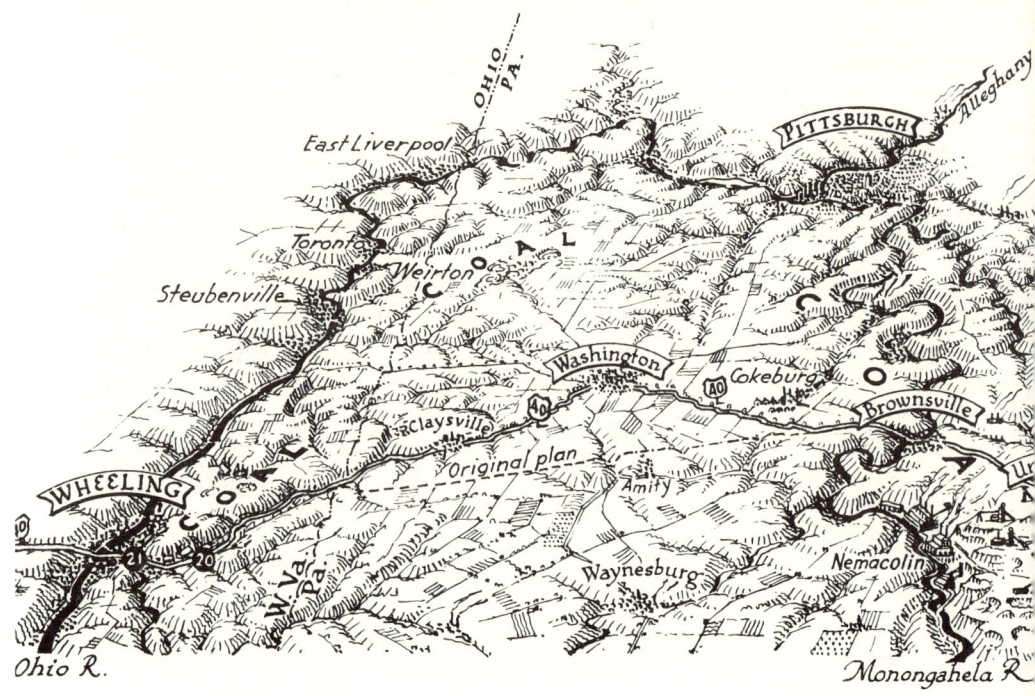

## NATIONAL ROAD

# Cumberland to Wheeling

From Cumberland, where it leaves the Potomac, to Wheeling, where it reaches the Ohio, a distance of 131 miles, U. S. 40 follows the route of the National Road—the Cumberland Road, as it was often called. This sector, like the original one from New Castle to Elkton, was thus in its beginnings a portage-route across a watershed. The scale, however, was about ten

*Surveying the National Road*

times as large. The Cumberland-Wheeling portage was over a hundred miles long, connected the waters of the Atlantic with those of the Gulf of Mexico, and avoided a difficult ocean- and river-voyage of several thousand miles. Moreover, instead of merely connecting two of the smallest colonies, it served to bind the whole nation closer.

Constitutionally, it set a precedent for the use of national funds for internal improvements. It contributed largely to the growth of Baltimore on the one hand, and on the other hand to the development of Ohio and the states beyond. In one way and another it may have done much to preserve the United States as one nation. No other section of the route, perhaps no section of any of our other highways, has been so closely connected with the great events of our history, or has surpassed this one for its actual influence upon the course of history.

Scenically, few sections of the entire route are more delightful. It passes first through the Allegheny Mountains, where the traveler encounters ridge-and-valley scenery much like that east of Cumberland, but on a larger scale. The general level is more elevated; the individual ridges, higher. The road crosses Negro Mountain at an elevation of 2908 feet, the highest point that it attains in the eastern United States. Tree-growth reflects the higher altitude, as dark stretches of pine appear among the lighter-colored deciduous trees.

Finally, the road crosses the Chestnut Ridge, and drops down at Uniontown to the definitely lower level of the Allegheny Plateau. This is no plateau in the sense of being level, except perhaps by comparison with the mountains to its eastward. Actually this southwestern corner of Pennsylvania, along with the West Virginia panhandle, is a maze of green hills—partially farmed, partially wooded, here and there blackened by the dumps of coal-mines. In contrast to the mountainous region—neatly laid out, as by a landscape-architect, in parallel ridges—the plateau seems a country lacking in structure, chiefly because it is without prominent natural features, except where the deep channels of the Monongahela and Ohio rivers cut across it. . . .

The origins can be traced back to 1748 when a group of Virginia gentlemen organized the Ohio Company, to exploit the Indian-trade in what was vaguely known as "the Ohio country." At that time the point in western Maryland at which Wills Creek flowed into the Potomac, where Cumberland was later to be founded, was the jumping-off-place. Even then the forested mountains to the northwest were not entirely unknown or trackless, and white hunters and traders had penetrated the region. Nevertheless, a practical pack-horse trail seems to have been lacking, and to explore such a route the Company delegated two redoubtable frontiersmen, Christopher Gist and Thomas Cresap. Gist himself had already journeyed across the mountains. As an additional aid the two employed a Delaware chief named Nemacolin.

In 1752 Gist and Nemacolin explored and blazed a trail from Wills Creek to the mouth of Redstone Creek, on the Monongahela River, the future site of Brownsville. Doubtless, when they could, they made use of the trails left by hunters and traders and even of ancient Indian paths. But the evidence, rather, shows that the pack-horse way was essentially something new. To maintain, because of one Indian hired as a guide, that the National Road stems from an "Indian path" is to quibble.

Nemacolin's Path, thus established, did not long retain that name or long remain unknown in history. In the winter of 1753-54 the twenty-one-year-old Major George Washington traveled over it, guided by Christopher Gist himself, on a mission to the French on the Ohio. In the following April Washington, now lieutenant-colonel, was again at Wills Creek, this time in command of some 150 Virginia troops, and entrusted with the task of forcibly, if necessary, ousting the French from the forks of the Ohio. The French, as it soon came to be known, were in possession, and fortifying themselves, and had also, according to reports that drifted back through the woods, a force much larger than the Virginians'. Still, as history was to show, George Washington was not a man easily frightened or turned from his purpose. There was good hope that he might be reinforced. On May 1, he began his advance, and thus made history, not only in general but also in the special field of highway history. For, though Nemacolin's Path was only designed for pack-horses, Washington set out to widen it for his supply-wagons.

The country was a continuous and thick forest; mountain ridge beyond mountain-ridge lay directly across the path; rain fell day after day, so that every flooded creek was a barrier. Four miles in a day was the best progress; sometimes it shrank to two; once there had to be a halt for two full days while the hard-working backwoods soldiers threw a bridge across a swollen stream.

Now and then an English trader, fleeing from before the French, came back along the trail, and told the dark wild rumors that fugitives tell—of French reinforcements pouring in from the north, of Indian scalping-parties already flung out far to the south.

But, to counteract this, better news came up from the rear, North Carolina was raising troops; so was Maryland. Pennsylvania, Quaker dominated, would not raise troops, but was sending ten thousand pounds; even far New

England was organizing to harry the French in Canada and divert their attention from the Ohio. At last the colonists were beginning to think and act together as a nation, and as the symbol of this new unity we may take that poor little narrow roadway, being pushed westward by the rain-drenched Virginians under the command of a young man named Washington.

The end, however, was not so pretty as it might have been.... Washington got his wagons ahead, over the mountains, for about fifty miles, to an open spot known as Great Meadows. He advanced with a few men—it would be called a patrol, these days. He surprised a little company of French, heard bullets whistle and boyishly thought the sound pleasant. He won a little victory in the skirmish.

Even so, the situation was bad. In expectation of the worst, Washington set his men to building a stockaded fort at Great Meadows, named Fort Necessity. Troops from New York were reported on the way, but the only reinforcements to arrive were three more weak companies of Virginians with some light cannon, and a hundred men from South Carolina, enough to bring the total to about four hundred.

Doggedly, Washington pushed his road ahead. It went up the hill beyond the meadow, and then swung off to the north, leaving the line of the present highway. The wagons and cannon were moved ahead. Then, at word of a French advance in force, there was nothing to do but retreat to the shelter of the little fort. The French closed in.

Outnumbered, in a bad position, drenched by continual downpours, the Virginians and South Carolinians fought staunchly. But casualties mounted, and after a while there was nothing more to be gained by fighting. Washington surrendered, and by the chivalrous conventions of eighteenth-century warfare was allowed to march home with military honors over the same road that he had cut through the forest....

In the history of our roads Washington's campaign is of importance. It showed the sceptics that the mountains—even the terrible Laurel Ridge—could be crossed by wheeled vehicles. For a year it was called Washington's Road. Then, as with Nemacolin's Path, Washington's Road was surpassed and renamed.

In 1755 the English tried again—this time with the 44th and 48th Regiments of Foot, with artillery, cavalry, and a naval detachment, *plus* colonial contingents, under the command of General Edward Braddock. Again the forces assembled at Wills Creek, where Fort Cumberland had now been erected.

Braddock set out to advance, but he was not content with the road as he found it—a narrow track, just wide enough for wagons to progress, in a pinch not more than six feet across. The general ordered widening to twelve feet, and in addition his standards of what constituted a road must have been much higher. He had little concern about the French, and considered the forest and the mountains his chief enemies. He felt, and most military men would agree, that the establishment of a good supply-line was imperative. So, comparatively speaking, he took his time about building the road. His engineering-officers relocated it as necessary, and really made it something new.

In the end, as every American may be expected to know, Braddock met disaster and death. Nevertheless his road remained, and since it was in most respects a new construction, it was properly known as Braddock's Road.

By whatever name, for several years it must have been almost untraveled. The French held Fort Duquesne and the English held Fort Cumberland, and in the woods no man's scalp was safe. Not until after the defeat of Pontiac's Indian confederates in 1763 was the country again moderately quiet.

In eight years many things happen to a hastily built road through forested mountain country. Slopes gully, hillsides slip, embankments wash away, bushes and saplings grow thick, trees fall. By 1763 Braddock's Road must have been little better than the original Nemacolin's Path, a mere pack-horse trail. Nevertheless, it remained the best road through that region—in fact, the only road.

This stage we find described, vividly and with charm, in Joseph Doddridge's *Notes on the Settlement and Indian Wars*. Doddridge wrote of the region which is now southwestern Pennsylvania, from 1763 to 1783. To him it was "Braddock's trail," and he never mentions it as traveled by wagons. Yearly, he tells, when the harvests were in and the approach of winter allayed the fear of Indian raids, little groups of settlers banded together and formed caravans of pack horses. From the outside world, they needed salt and articles of iron and steel. In return, they had furs to barter. At first the pack horses, each with his bell a-jingle, had to be led all the way to Baltimore before a market could be found. But it moved steadily west—to Frederick, Hagerstown, Oldtown, and finally Cumberland. Except for Oldtown, each of the points marks a progression of the market-center westward, along present U. S. 40.

Thus the situation remained, essentially, for a generation or more, along the old route. A better road led across the mountains farther north from Philadelphia, and at its western terminus Pittsburgh grew up as the first western city. There was no important change until after the beginning of the new century.

Jefferson was president. Always restless-minded, always nationally oriented, he had looked to the west and purchased Louisiana. Ohio had been admitted. Yet there was always a chance that Ohio, along with the other western states, might slip away from its union with the older states on the Atlantic seaboard. The acquisition of Louisiana perhaps even increased this possibility, for now the westerners had free access to the mouth of the Mississippi over the unexcelled system of inland waterways offered by that river and its tributaries.

There is no need to go back into complicated matters involving constitutional interpretation and into finances. We may start with the report of a Senate committee on December 19, 1805. The committee's problem, which went back to the terms of agreement under which Ohio had been admitted to the Union, was to recommend the termini of a road "from the navigable waters emptying into the Atlantic to the river Ohio, to said State." They rejected a northern route leading from Philadelphia, and a southern one leading from Richmond. Another route led westward either from Baltimore or from Washington. The committee favored this central route, but recommended that the road begin at Cumberland, considering that the Potomac was potentially, if not actually, navigable as far up as that point. They recommended a western terminus on the Ohio somewhere between Steubenville and Wheeling. Echoing the words of the prophet Isaiah, the committee closed on a high note:

> to make the crooked ways straight, and the rough ways smooth will, in effect, remove the intervening mountains, and by facilitating the intercourse of our Western brethren with those on the Atlantic, substan-

tially unite them in interest, which, the committee believe, is the most effectual cement of union applicable to the human race.

Nonetheless, there must have been many who said that the committee was acting from motives far less high than their words indicated. To Philadelphia and Richmond alike, the location of the new road represented rank favoritism for Baltimore, and in the final vote of March 22, 1806, the Pennsylvanians in the House voted 13 to 4 against the measure, and the Virginians joined them by 16 to 2. The bill passed by only 66 to 55.

Although the road was never specifically named, it was designated in the first bill as "a Road from Cumberland," and it soon came to be known, officially and unofficially, as The Cumberland Road. Later usage has preferred The National Road, partly because the other term is somewhat ambiguous, suggesting the route through Cumberland Gap, which is actually the famous Wilderness Road.

In the original act the locating and preliminary surveying were entrusted to "three discreet and disinterested citizens" to be appointed by the President. These commissioners, duly appointed, having hired a surveyor, two chain-carriers, a marker, a vane man, and a pack-horse man with his horse, met at Cumberland on September 3, 1806, and set out on their exploration. The necessity of using a pack horse is evidence of the roadless nature of the country, even after 1800. In their report, dated December 30, they recommended Wheeling as the western terminus, because the river was more easily navigable below that point and because that town lay more directly upon a western course. They also established a route from Cumberland to Wheeling, although—as they reported a little dourly—their task had been rendered more difficult "by the solicitude and importunities of the inhabitants of every part of the district, who severally conceived their grounds entitled to a preference."

Before the road could actually be constructed, approval was necessary by the legislatures of each of the three states through which it passed. Maryland and Virginia gave ready consent, but Pennsylvania demurred, not altogether without reason. The proposed road, 112 miles long altogether, would pass for 75.5 miles through Pennsylvania. Being populous and wealthy, that state would pay heavily through national taxes, and yet would gain little, and might well lose heavily. Moreover, of the road-commissioners, two had been from Maryland, one from Ohio, and none from Pennsylvania at all. The road had even been laid out so as to by-pass both the county-seats of Pennsylvania that lay near its course. By and large, therefore, it is surprising that Pennsylvania did not raise more objection than it did. In the end its legislature approved the road, but added the proviso that it must be routed through the two county-seats, Uniontown and Washington.

The first of these raised little difficulty. In fact it lies so directly upon the straight line between Cumberland and Wheeling that one is a little suspicious as to why it was by-passed in the beginning. Washington, however, lay definitely to the north. Nevertheless, the road was so constructed, and a distinct northern tip on U.S. 40 still preserves the visible evidence of this ancient rivalry.

As thus approved, the route followed the general line of Nemacolin's Path as far as Brownsville, and of Washington's and Braddock's roads as far as the point where they turned off definitely to the north—which happened to be the place at which the General himself was buried. There was, naturally, much relocation, and as far as the crossing of the Youghiogheny, according to the statement of the commissioners, the new road did not follow the old one at any part for more than half a mile or for more than two or three miles altogether.

There had to be more detailed surveys and further acts of Congress, and the first actual contracts were not let until May 8, 1811. By the end of 1813 only ten miles of the road had been opened. It was an oxen-horses-mules, ax-pick-shovel job. The building of bridges took time, and special road-working machinery had not been invented. No one in America knew much about road-engineering. Congress hesitated and virtually trembled every time it appropriated amounts as large as $100,000. Nevertheless, the road reached Wheeling in 1818.

It was a notable achievement. . . . A strip 66 feet wide, cleared of trees and underbrush. A leveled roadbed thirty feet wide. Grades and curves moderate, so moderate indeed that many of them still remain on the present highway. Twenty feet of the surface covered with broken stone from a depth of twelve to eighteen inches.

No wonder that an old-timer like Doddridge, writing not long after its completion, contrasted the "horse paths" of his youth with the new road which had opened up "the distant region not many years ago denominated the backwoods," and connected it with the eastern cities "as if by magic enchantment."

Traffic poured over the new highway—stage-coaches with passengers and mail, "movers" with their household goods in farm-wagons. But most of the tonnage moved in the great freighting-wagons, some carrying a "hundred-hundred," or five tons, called Conestoga wagons after the Pennsylvania valley where they were first manufactured. Inns and taverns, blacksmith-shops, whole new towns—sprang up along the road. Old towns boomed and flourished.

According to an Ohio congressman's statement, in 1822 one mercantile house in Wheeling unloaded 1081 wagons, averaging about 3500 pounds of merchandise. He estimated that altogether about 4700 wagons had arrived that year in Wheeling, consigned to regular commission houses. About one tenth as many wagons were estimated as passing through Wheeling to western points, and the business of transshipment at Brownsville, head of navigation on the Monongahela, was roughly figured at two fifths of the whole. The organized freighting business of this year should therefore total about 8500 wagons, carrying 15,000 tons. In addition there would be a considerable amount carried by stage-coaches, by "movers," and by wagons not consigned to any commission house, and the whole figure must be doubled, or perhaps more than doubled, for in all probability the flow of tonnage eastward was greater than the flow westward.

Although the road had been constructed for wheeled vehicles, it turned out to be a great drovers' road. In fact a considerable proportion of the eastward-moving tonnage traveled on its own feet—horses, mules, cattle, sheep, hogs, even turkeys. It is reported that geese were so driven, being now and then made to walk over soft tar to supply them with a protective coating for their feet.

Sad to say, there was also a thriving traffic in slaves, from the now declining agricultural areas of Maryland and Virginia, where more slaves were produced than were needed, to Wheeling, where they could be "sold down the river" to the booming cotton-country. Thomas Searight, first chronicler of what he calls "the old Pike," records that he himself had seen them "driven over the road arranged in couples and fastened to a long, thick rope or cable, like horses."

The Cumberland Road, like all others, had scarcely been completed before it began to deteriorate. Then, by some curious legalistic twist, doubts arose. Congress might have the power to construct a road but not to maintain it. President Monroe, in 1822, vetoed a bill for the establishment of tollhouses to provide funds for repair, even though, aside from the question of constitutionality, he approved. The repair of the road became a political issue. In one way or another funds were sometimes appropriated for that purpose, but they were generally inadequate.

A rather ignominious end came to the first national experiment at highway construction when in the 1830's the road was handed back to the states through which it passed, and these states erected their own tollhouses. The road thus became a turnpike, and was often called the National Pike. By that time it had already passed its heyday, and the construction of the Baltimore & Ohio Railroad had begun. The completion of that railroad to Wheeling in 1852 finally ended the great period of the National Road.

It had been well laid out and constructed in the beginning, and its tollgates continued to bring in some revenue for its maintenance. Even through the Dark Ages it was always much better than the average American highway.

In 1915 a gentleman with the historic name of Robert Bruce drove along the road, and in the next year published *The National Road*, a small but excellent volume dealing both with past and present, amazingly well illustrated by means of photographs and detailed maps. His pictures show the road, although unpaved, to be in good condition. Some of my own photographs were taken from the same spots as some of those reproduced in Mr. Bruce's book, and except for the pavement show almost no changes as far as the road is concerned.

Since that time the road has been paved and in some places widened and straightened. The Pennsylvania Department of Highways estimates that fifty per cent of U. S. 40 within that state has been relocated, for one reason or another. Many of these relocations, however, have not moved the road very far, and in Maryland there has been much less change. On the whole, U. S. 40 still follows very closely the line of the National Road.

The northern detour by way of Washington has not, however, been eliminated. This adds a distance of nearly twenty miles, for we may remember that the road, as originally surveyed, was to be only 112 miles.

The future will almost certainly produce further relocations. The whole northern angle with its apex at Washington may at some time well be eliminated. But the original location was so good and the route so often approaches the ultimate perfection of the straight line that one may expect indefinitely to continue driving over much of the same actual ground once traversed by the Conestoga wagons.

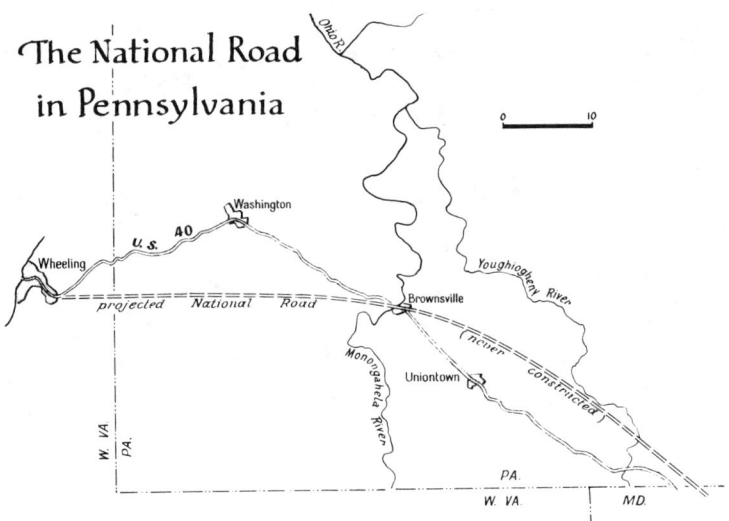

# ⓴ The Narrows

🛣️ To right and to left of the highway, just north of Cumberland, rises Wills Mountain, a typical Appalachian ridge. Though cut squarely in two by Wills Creek, it rather curiously maintains the same name on both sides, as if the roots of the mountain joined beneath the running water, or else as if even the early pioneers who fixed the name recognized the geological unity of the two halves, accidentally divided, as it were, by the creek. This is what is known as a "water gap," cut downward by the stream, widened at the top by rain and frost, so that the result is a broad V-shaped passageway.

We are here, as is indicated by the higher ridges in the distance, just at the beginning of the Allegheny Mountains. Wills Mountain itself rises only to heights between 1600 and 1700 feet, but the high ridge obscured by clouds in the left background is Piney Mountain (ca. 2400), its name an indication that it rises high enough to be in the pine-belt.

Looking into the Narrows, more than five hundred feet down, the observer sees the way in which the routes of transportation have crowded into this narrow break. In addition to the highway there are five lines of railroad track, four pole lines, and a rough road, scarcely more than a track, that parallels the stream and crosses it by a ford.

Rather curiously, the first route was not through the Narrows. Washington, probably following Nemacolin's Path, took his road over Wills Mountain, to the left. Braddock's engineers started to widen Washington's road, but found the work extremely difficult. They did not know of the easier way, and apparently everyone else in the army, including Washington, was ignorant also. The actual discoverer of the route was the enterprising Lieutenant Spendlowe, in command of the detachment of sailors.

Oddly, the engineers of the National Road preferred Washington's route, and their original road was so constructed. In the end, however, it proved too difficult, and the only major relocation of the old National Road occurred when it was constructed through the Narrows, just before being turned over to the state. Thus, getting there first, the highway took, and retained, the better location. The railroads have been forced to dig into the hill at a higher level.

The ridge in the middle distance shows one or two open fields, probably pastures, and an old orchard. Like so much of the mountain country, however, it is returning to woodland. A picture taken about 1915 shows that most of its slopes were then clear of trees.

Although cloud and mist blot out much of the farther view, the effect is characteristic of the Appalachian country, where mist and mingled coal-smoke are generally a part of the scene. One result is that distances are made to appear greater. Piney Mountain is actually only three miles distant, and even Little Allegheny Mountain, the farthest ridge, is but four miles away.

## ⓯ From Little Savage Mountain

🛣️40 The view westward from the slope of Little Savage Mountain scarcely gives the impression that one is at an altitude of 2700 feet, and well up toward the top of the Alleghenies. The level-topped ridges rise only a few hundred feet above the general plateau-level and succeed one another at such even height that they hide the intervening valleys and produce an impression of level or gently rolling country. Notable in the distance at the left is the long ridge of Meadow Mountain, slightly under 3000 feet, falling off at the right to a water-gap.

The little stream at the bottom of the valley flows southward to the Potomac and the Atlantic Ocean, and U. S. 40, swinging toward the left remains in that watershed for several miles more. Actually, however, because of the curious interlocking of stream-headwaters in the Alleghenies, one can here cross into the Mississippi watershed by merely following up the valley to the right a short distance and going over an insignificant rise.

As elsewhere in the Appalachians the ridges are generally wooded, now chiefly with oak, as exemplified in the tree that overhangs the highway.

The highway here follows exactly on the line of the National Road. Just beyond the farthest barn, where a white truck shows, begins what was known in the great days of the road as "the long stretch." This is still, up and down grades, a straightaway of two and a half miles. It was renowned as the longest straight stretch on the whole original road from Cumberland to Wheeling.

A marker declares this to be Braddock's third, or Savage River, camp. It was located probably in the meadow, some distance to the left of the highway.

After the building of the National Road a tavern was located here as early as 1830. It was kept by Thomas Johnson, who was noted as a fiddler and owned a slave famous as a dancer of the double shuffle, so that master and man frequently performed for the entertainment of the customers. A tradition of boisterous jollity prevailed in the taverns along the national road, and the consumption of whiskey was high. The present frame house, however, with its mansard roof cannot date from much earlier than the sixties.

At present the farm, like so many in the eastern countryside, has a city owner. It is given over to stock-raising, chiefly to horses. Its cleanly painted white fences are in fact almost an advertisement that it is so devoted. The old apple-orchard behind the house has been allowed to deteriorate, and trees have died out until the actual pattern of an orchard is scarcely distinguishable.

The arrangement of the farm with the barns across the road from the house is an interesting survival of a quieter period. It was a practical arrangement before the automobile era, and is very commonly found. With the development of heavy and swift-moving traffic it becomes an inconvenience and even a hazard, but it nevertheless still survives.

Bruce's *National Road,* published in 1916, contains a picture taken from almost exactly this same spot. It shows no change in the alignment of the road at all, and very few changes of any kind. Even the fence in the foreground with its out-of-line posts and old sagging wire is apparently the same.

**16** Mason-Dixon Line

(US 40) U. S. 40 winds among the hills, curves toward the northwest, and crosses from Maryland into Pennsylvania. Halfway up a little hill at the state line, the surface of the highway changes from black to white, and thus gives visible and concrete reality to one of the most famous abstractions in American history. This is the Mason-Dixon Line.

Known by the names of the two surveyors who determined it in 1767, the line settled the long disputed boundary between the colonies of Maryland and Pennsylvania. In pre-Civil-War days it became the symbolic division between slave and free territory, and most Americans think of it in this secondary rather than in its primary meaning.

For complete accuracy the line should not cross the road at right angles, but should run at an angle bringing its right-hand end much nearer the observer than the end at the other side of the road. Apparently, however, road-making machines are not adjusted for working at such close and accurate angles. Maryland here has given Pennsylvania a certain benefit of the doubt, or else the Pennsylvania roadmakers have not worked down quite to the Maryland line.

Another interesting though common feature of the road on this section is the raw, ungrown cut at the side of the road as it ascends the hill. We are used to such raw cuts and take them for granted. Yet actually they are a symbol of our changing civilization, and our grandchildren may not know them. Such a cut will grow over in the course of a few years, and no longer be so sharply distinguishable from the rest of the landscape. In the eastern part of the United States, if roadwork ceased for as long as ten years, the raw cut would largely disappear from the landscape. A growth of bushes and saplings is already making headway on this cut.

The scene is here typical of much of our eastern hill country. The hardwood forest is encroaching upon what was once a farm. We can still see an old barn, a few trees of an orchard, and some open fields that may once have been plowland and now have reverted to nothing better than rough pasture at most. The edges of the field, moreover, have grown fuzzy as the trees have started to take over.

The little American Gothic building at the left of the highway is the State Line Methodist Church. Like the abandoned field it also bespeaks a time of thicker inhabitation.

The three identical cars on the highway belong to a company fleet, and are probably being driven eastward from a mid-western factory.

## ⓱ Fort Necessity

🛣️ The view is from inside the reconstructed stockade bearing the name of Washington's original Fort Necessity. Halfway up the side of the hill the line of modern U. S. 40 can be made out.

This neat National Monument gives little of the effect of the hastily built, irregular quadrangle of the original fort. Indeed it is almost the antithesis of what was here on that miserable day of battle—a trampled quagmire of mud under the never-ceasing downpour, corpses sprawled, wounded men huddled about, dead horses and dogs, and still the drenched and desperate Colonials keeping up the fire under the command of that harassed young Virginian, whom nobody as yet was calling the Father of his Country.

Yet, if Fort Necessity National Battlefield Site does not supply us with any close facsimile of July 3, 1754, it nevertheless is a good replica of a frontier stockade and a log cabin. The posts are about seven feet high, with rifle-slots cut down to firing-height. The total height of the palisade has been increased by having the posts rise from a raised ridge of earth. On the inside, this ridge serves as a firing-step.

Although there is no certainty that any log cabin stood within the confines of the original fort, the present one is a good reproduction. Its saw-cut logs around the windows and doors, its also saw-cut door-jambs, and its numerous windows are not typical of the early cabins. On the other hand, it is structurally independent of nails, which were commonly not obtainable on the frontier. The log-ends have been notched and crossed, so that each log holds the lower one firmly in place. The space between has been chinked—in this case cement has been used, but mud would originally have served. The roof consists of shakes, that is, long shingles split naturally with an axe from some straight-grained log, held in place without nails by means of wedged crosslogs.

At the time of the battle, thick forest covered all of the hillsides, but the meadow was too wet for tree growth. The Colonials had a field of fire of not more than about a hundred yards, and the French and Indians therefore had the shelter of the trees. The position was really a trap, and only an inexperienced officer would have tried to defend it against a superior force. The French probably advanced at first from the northwest where the cornfield and the young trees now stand. Later they attacked from other sides as well.

The background of the picture shows the typical rolling country of western Pennsylvania. Theoretically such a hillside should never be planted in corn, but the heavy soil of that region does not erode easily, and even after a period of cultivation which may extend back for more than a century, the field is producing well. The ample barns are another indication of a good farm.

The hill at the right is known as Mt. Washington, and the brick building is the Mt. Washington Tavern, built about 1818, as a stage station on the National Road. With its simple lines and its double end-chimneys it is a good example of the better-built taverns of that time.

## ⑱ Braddock's Grave

(40) Edward Braddock, like his father before him, had served for many years as an officer of the Coldstream Guards. He was a run-of-the-mill professional soldier, having an undistinguished record behind him. His appointment, with the title of Major General, to command the expedition against the French at Fort Duquesne is somewhat inexplicable, unless we merely consider that the high command did not consider the campaign to be of any great difficulty.

Braddock was a disciplinarian of the damn-'em-and-flog-'em school, personally brave, as most such officers are. He misconceived the problem of fighting in forest country, and suffered ignominious and bloody defeat. He himself was shot through the body—as a frontier legend declared, by an American whose brother the general had too severely disciplined. More likely he was shot by a French bullet. In any case, he was carried back with the retreat, and died near the spot where his monument now stands.

The picture itself is not inharmonious with the mournful and inglorious death—the dark tree, the hills muted by mist, the low-hanging gray cloud, the crucifix-like pole and cross-arm. Only the gay flowers of the foreground yield an incongruous touch of gaiety.

At the left, beside the highway itself, an official marker of the state of Pennsylvania declares this to be Braddock Park. The roads of Pennsylvania are probably better supplied with historical markers than those of any other state along U.S. 40.

The bronze plaque, set in the stone at the extreme right, marks "The Old Braddock Road." By this is meant, not the modern track in the foreground, but the trace showing as a swale just to the left of the marker. Natural roads tend to wear down below the general surface of the ground, as the result of the pulverization of the soil under traffic and subsequent work of water and wind. The plaque summarizes, not without eloquence, the history of the road. One must remember, as the tablet emphasizes, that Braddock's road remained in use for about seventy years; pack trains as well as cannon followed the route.

The tablets on the tomb itself record something of Braddock. He died on the night of July 13, and was buried the next morning, "decently but privately," in a grave dug in the middle of the road. The surface was then leveled off, and the troops were marched along the road and the wagons driven over it, to hide the grave and save it from being desecrated by the Indians.

In 1812, fifty-seven years later, a skeleton with military trappings was unearthed in the old road-bed, and was later removed to the site now marked. Its actual identification with Braddock seems to be a reasonable likelihood. The monument was later erected, with an officer of the Coldstreams imported to represent the old regiment.

Apparently the approach of death enabled even the stodgy Braddock to speak with a little of the poetry that we associate with the death of a tragic hero. "Who would have thought it?" he muttered as they carried him back along the road. As he lay dying, he said with infinite unconscious irony, and yet with accurate prophecy, "We shall better know how to deal with them another time!"

# ⓳ Toll House

🛣️ An excellent marker supplied by the state of Pennsylvania declares this to be "One of the six original toll houses on the Cumberland or National Road. It was built by the State after the road was turned over to it by the U.S. in 1835. The road was completed through this section in 1817-18."

The Historical Society of Western Pennsylvania declares, with chagrin, that it is unable to give details of the architectural origins of this tollhouse. In its brickwork, simple stone lintels, and general restraint, it follows obviously in the Georgian tradition, which was still strong at the time of its building in spite of the advances of the Greek Revival. Its curious features, however, are its octagonal, tower-like shape, with the center chimney or stovepipe-outlet. Probably these features are to be explained on the ground of economy and utility. To enclose a given number of square feet an octagonal shape demands less material than a square one, and this factor was of some importance when the building was being carried up to a second story. The height was probably to give the tollhouse-keeper a long view in both directions. One may note that the building is liberally supplied with windows, although some of them have been bricked in.

There have been some attempts at preservation—observe the new roof on the tower—but the building is in danger of falling to pieces. Much of the glass is gone. The porch-roof and the roof of the rear extension have both broken. The porch posts at the front, where doubtless they were occasionally banged by wagon hubs, have gone out entirely.

The National Road probably went by at close to the level of the porch. The retaining-wall indicates that sometime in the early automobile era, probably thirty years or more ago, the top of the hill was cut off to attain an easier grade and better vision. The stretch of road here in view, with its roller-coaster effect, shows the way in which the engineers of the National Road sacrificed easy grades to save distance, and so produced many straight stretches, even across rough country. Since a horsedrawn freight-wagon moved at a walk, a mile of distance which could be saved compensated for a considerable amount of uphill and downhill.

The view here is east across the rolling country of the Allegheny Plateau. The high and level skyline of the Chestnut Ridge stands out faintly in the distance, the first of the mountains, as one comes from the west.

The prominence of billboards indicates that the location is only about two miles from the city of Uniontown. A telephone line follows the highway at the left, and at the right rises the L-shaped pole carrying the three wires of a small local power line.

Note the roof of a car rising above the line of the highway in the approaching lane. It has a little of the look of a crocodile's head rising above the water level, but it is much more dangerous. Having taken the picture, the photographer had about a second to gather up his tripod and get out of the way.

## ⓴ Coal Mine

🛣️ At best there is little that is picturesque about a coal mine, and when a leaden sky merely echoes back the grayness of the dump, picturesqueness is at a minimum. Yet U. S. 40 runs through coal-mining country for 150 miles, and this basic industry can scarcely be omitted from any presentation of a cross section.

The location is about halfway across the northern panhandle of West Virginia, through which U. S. 40 cuts for a few miles. Little Wheeling Creek is here busily engaged in carving its valley down through the Allegheny Plateau. The raw exposures of Carboniferous sandstones and shales, however, are the immediate result of road-building at the left and railroad-building in the middle center, not of stream cutting. Seen thus from a lower level, the plateau-country seems quite without design. Actually, however, the hilltops here are all of about the same height, representing the level of the elevated plain from which the stream valleys have been cut down. The hill at the right merely appears to be higher because it is closer.

The vast dingy-gray dumpheap of the mine occupies most of the foreground. At the right, served by a siding, are the works above the mine-shaft where the crude coal is brought up from the shaft, worked over, and loaded into cars. These buildings with their wholly functional design actually reproduce some of the effects of slanted planes at which modern architects labor.

On the hillside, startlingly white, stand eight double houses, constituting the company town. It is a better built company town than many in the coal-mining regions, and its outhouses are decently obscured among the growing trees.

At the left the picture shows not only another stretch of the original National Road, but also a little of the Baltimore & Ohio Railroad, which killed the National Road.

Both highway- and telephone-engineers have barely avoided a shudder when looking at this picture. The three-lane highway is considered by modern standards to be a deathtrap. It usually indicates that a hard-pressed highway commission, limited in funds, was desperately striving to increase the carrying capacity of a road. In this instance, as is common, the three lanes are reduced to two at the rounding of the curve.

"Boy!" was the remark of a telephone-engineer at noticing the pole with eight crossbars. "You don't see those things often these days!" The amount of trouble that must arise when an ice-storm strikes such a line of poles is frightening to contemplate.

The peach trees rising in the center of the picture are the planting of an enterprising Negro farmer who has occupied a small hilltop. Only one little field shows in the picture, though the thinness of tree-growth elsewhere indicates that the forest may just be engaged in taking over some old pastureland. Across the Appalachian country the forest now is probably advancing much more rapidly than it is being cleared away.

# ㉑ Wheeling

🛣️40 Looking out across the rooftops of Wheeling, West Virginia, the observer stands on the hillside about a thousand feet above sea-level, and about four hundred feet above the main channel of the Ohio River, its glassy surface unruffled by either wind or current.

The country is here the Allegheny Plateau, at this point deeply channeled by a major stream. The land just beyond the water is Wheeling Island, part of West Virginia and separated from Ohio by a smaller channel of the river. In the distance are the hills of Ohio, dissected by the valley of Wheeling Creek. The apparently level skyline represents the old peneplain, from which the rugged plateau has been carved. The skyline-level is between twelve and thirteen hundred feet, and is deceptive in that it does not represent a continuous north-and-south ridge, but is merely a distant conglomeration of smaller hills and ridges. Actually, as in the preceding picture, the hill that looks higher than the distant skyline merely appears so, because seen closer to, from a lower level.

In the foreground is a part of the flat-roofed business district of the city, with a few old houses along the riverfront.

The cramped location of Wheeling, crowded between hill and river, is well shown. This was an important reason why the city, in spite of being the terminus of the National Road and of the Baltimore & Ohio Railroad, was never able to rival Pittsburgh. Expansion onto the island was difficult because of the danger of floods.

An interesting historical reminder of the National Road is to be seen in the factory sign, which must be read backwards, MARSH. This company, founded in 1840, still manufactures the kind of cigars known as stogies, which were favored by teamsters and were originally known as Conestogas, taking their name from the wagons operating on the National Road.

The nearer bridge is interesting not only as an excellent example of an early suspension-bridge, but also for its historical association with the route of U. S. 40. In 1849 a suspension bridge was built across the river at this point, much against the wishes of the citizens of Pittsburgh, who feared that a bridge would interfere with navigation and thus make Wheeling the chief city on the river. This bridge was blown down in 1854, and a case was then carried to the Supreme Court as to whether another could be constructed. Wheeling won, and the present bridge was completed in 1856. At that time it was the longest suspension bridge in the country, and its 1010-foot span still takes the traffic of U. S. 40 up to the six-ton limit.

This picture represents the end of the original National Road, and the beginning of its later extension.

The route of U. S. 40 can be followed in the picture. After crossing the suspension-bridge, it swings diagonally across Wheeling Island to the other bridge, following the line of Zane Street. The modern street not only bears the name of Ebenezer Zane, but also must follow rather closely the line of his trace. The highway then proceeds westward, up the valley of Wheeling Creek and along the route of the National Road, as extended, which is also the general route of Zane's Trace. In ten miles the highway has reached the height of 1260 feet at St. Clairsville, and has attained the level of the skyline, as here seen.

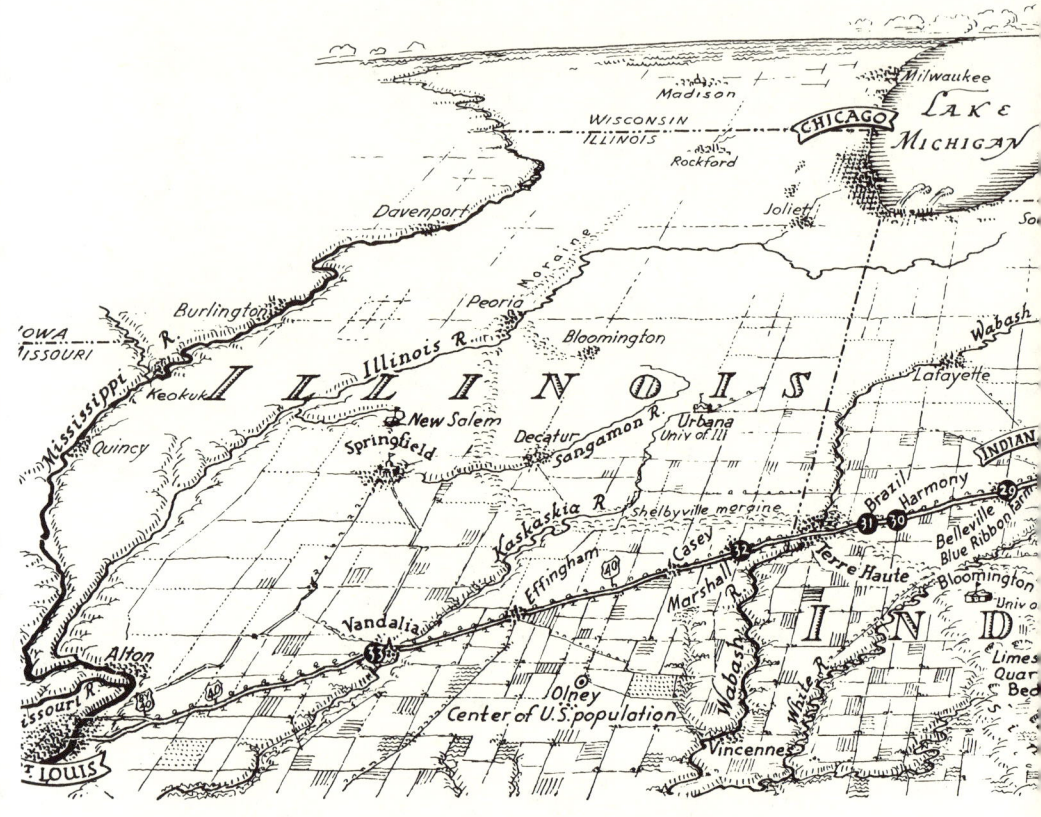

## NATIONAL ROAD, EXTENDED
# Wheeling to St. Louis

**U.S.** 40 across Ohio, Indiana, and Illinois may be considered a unit, even though its length of 531 miles is greater than that of all the four eastern sections combined. It extends from the Ohio River to the Mississippi. Topographically, it first crosses a hundred-mile stretch of the Allegheny Plateau, and then passes for nearly four hundred miles over the level plains

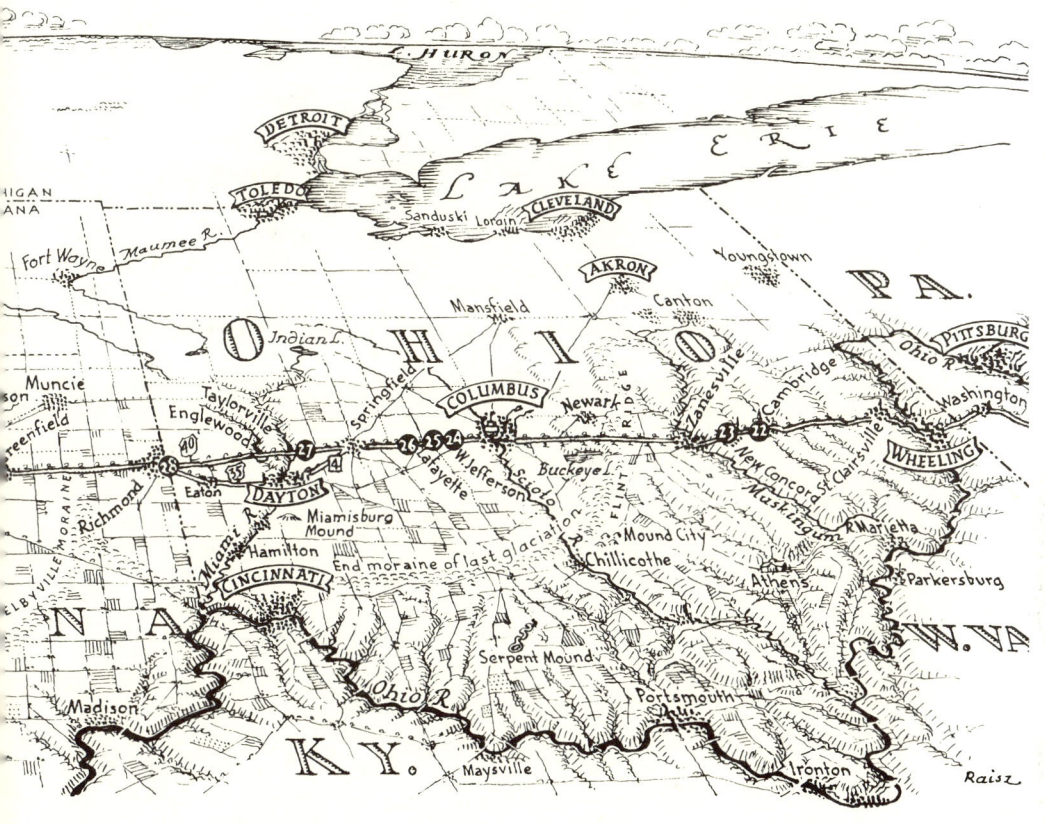

of the Central Lowland. Historically, except for the few miles west of Vandalia, Illinois, it represents the extension of the National Road.

To the ordinary tourist's eye this run offers little of scenic interest. Eastern Ohio is pretty enough—rolling hill-country of woods and pastures. Farther west, however, the flatland presents nothing that is conventionally picturesque. The attention here should rather focus upon the well-tended and rich farms of the Little Corn Belt, and upon the tree-shaded firmly-founded small towns, most of which have already celebrated their centenaries. Columbus and Indianapolis, on the other hand, present little that is distinctive among American cities, and have not even provided by-passes to keep the highway from plowing drably through them, along mile after mile of secondary business-districts.

Primevally, thick forest extended, almost without openings, as far as the western edge of Indiana. From there on, the country was different, offering the sharp contrast and the scenic variety between woodland and prairie. All this has disappeared. The settlers cut the forests, and planted trees around their houses and barns in the prairies. As a result, only a well-trained observer can tell from the look of the country, chiefly by the darker soil and richer farmlands, when he is passing across an originally treeless prairie.

This sector springs from no historical routes—or roots—older than the National Road, except for the first seventy-three miles west from Wheeling. Here we must go back to Ebenezer Zane, Virginia-born frontiersman, back-woods-storekeeper, Indian fighter—a man of no mean abilities. Having blazed a trail from Pittsburgh to Wheeling, he began to have further ideas. In 1796 he petitioned Congress for the grant of three tracts, each one a mile square, in return for which he would complete a road that he had already begun and would establish ferries at its three principal river-crossings. The petition was granted, and Zane carried out his end of the bargain.

His road, known as Zane's Trace, left Wheeling in a westward direction, and then, swinging off to the southwest, ended at the Ohio River opposite the town of Maysville, Kentucky. Only its first section, as far as the Muskingum River, is involved with the history of U. S. 40. At this river Zane established the most easterly of his three ferries, and here he located one of his mile-square tracts, where now is the city that aptly bears his name.

In the summer of 1796 Zane began work on the road with a party of six or eight men. They blazed trees, cleared out the thick underbrush, and removed fallen tree-trunks. They had pack-horses with a tent and provisions, but they lived largely on game. Two men kept watch at night, for there was still some fear of Indians, in spite of their defeat at the Fallen Timbers in 1794.

The trail-makers followed the course of Wheeling Creek for about seven miles. After that they took the road up to the ridge, and kept on westward, generally avoiding marshy lands and keeping high, after the manner of Indian trails. In some places they may actually have followed the old Mingo Trail. In general they headed west, and even the modern road has not departed very far from their course.

They opened only a horse-trail. In 1804, however, the Ohio legislature passed a bill for a wagon-road, appropriating $15.00 a mile for this purpose—by modern standards probably not enough to pay for the postage stamps expended in the process of constructing the same length of four-lane highway.

The road that was opened as the result of so limited an expenditure was naturally not outstanding. Probably the trail was widened, straightened where necessary, relocated in spots to ease the grades a little, and dug out on the steeper side-hills to keep the wagons from tipping over. There would have been no attempt at surfacing, most likely. Stumps were not grubbed out, but were left standing to a height of fifteen inches, which wagon-axles would clear—if the ruts were not deep. A "mover" has recorded that in 1806 he took two days to get his three wagons up from Wheeling Creek to the top of the hill at St. Clairsville, about four miles. Zane's Trace thus stands to the westward extension of the National Road much as Braddock's Road stands to the original road farther east.

In 1820, only two years after the Conestogas had begun to roll into Wheeling, Congress appropriated money for further surveying. To avoid such difficulties as had produced the Washington detour, the law required the road to run in as straight a line as practicable. Its western terminus was to be a point on the Mississippi River "between St. Louis and the mouth of the Illinois." Even this slight vagueness was to have unfortunate political results.

A straight line within the limits specified was found to send the road south of both Columbus and Indianapolis, although the linking of state capitals was a basic principle. As a result the law was modified in 1825, to let the road pass through those two cities and also through the newly established Vandalia, capital of Illinois, and to extend the survey as far as Jefferson City, capital of Missouri. Actually the changed law did not make the road depart greatly from straightness, because Columbus and Indianapolis happened to line up approximately with Wheeling.

The survey progressed rapidly—as far as Columbus in 1825, to the Illinois line in 1827. By 1829 preliminary surveys had been completed, with alternate northern and southern routes from Vandalia to Jefferson City, waiting upon a decision as to whether the Mississippi crossing should be made at St. Louis or at Alton.

Real construction was much slower. Ground was broken at St. Clairsville, Ohio, in 1825. A prayer was offered, the Declaration of Independence read, and an oration delivered. The orator declared that the road was destined to reach the Rocky Mountains.

In 1830 the road was completed to Zanesville; in 1833, to Columbus; in 1838, to Springfield. But beyond that point its continuity began to break down. Much work was done, indeed, not only in western Ohio but also in Indiana and Illinois. The road was pretty well "grubbed, graded, and bridged" clear across Indiana, but only a few miles, in four isolated segments, were finished. In Illinois, as far as Vandalia, the right-of-way was established, and the clearing was done after a fashion. But that was about all, and even the stumps were mostly left standing.

This gradual breakdown of construction is partly attributable to the continually growing interest in railroads. In 1836 Congress seriously considered the financing of a railroad instead of a highway, west of Columbus. Partly, however, the extension of the road was hindered by local rivalries which made states and even towns fight so viciously that in the end they prevented the road being built by either route.

One such war-to-the-knife was waged by the towns of Dayton and Eaton in Ohio. The state legislature supported the two towns, but they lay a little off the direct route, and the road eventually by-passed them both, as U. S. 40 still does. In retaliation, or self-defence, the citizens of the towns backed a privately financed turnpike which diverged from the National Road at Springfield and matched that road in every detail, even to having mile-posts showing the distance from Cumberland. As a result there was really no need to finish the National Road from Springfield to the Indiana line. The Dayton-Eaton turnpike still exists as parts of Ohio 4 and U. S. 35, and since Dayton has grown to be a large city, this route is partly four-laned and remains in this area a more important highway than U. S. 40.

Another curious reminder of this controversy is the name of the village of Vandalia, Ohio, some twenty miles west of Springfield. When it appeared that the National Road would never reach the capital of Illinois, the Ohio settlement took the name, thus giving a new twist to the story of Mahomet and the mountain.

More disastrous and less humorous was the struggle between Illinois and Missouri. Illinois fought for Alton, and Missouri fought for St. Louis, and each state fought so desperately that the road was never authorized beyond Vandalia at all.

In this manner petering out toward the west, the extension of the National Road never attained the greatness of the original. It served the central parts of Ohio and Indiana, rather than the nation as a whole. According to Searight, who wrote largely from his own recollections, "The travel west of Wheeling was chiefly local, and the road presented scarcely a tithe of the thrift, push, whirl and excitement which characterized it, east of that point."

But if it lacked the long files of Conestoga wagons, it had its stagecoaches with passengers and mails, and was an important road for western emigration—for "movers" with their wagons and domestic animals. For, even though it was not much of a road, it was the best road there was! It brought to large areas of three great states some touch with civilization, two decades before the coming of the railroads.

To Benjamin S. Parker, who as a child lived near the road in Indiana, it meant much:

> With the tinkling of the bells, the rumbling of the wheels, the noise of the animals, and the chatter of the people, as they went forever forward, the little boy who had gone down to the road from his lonesome home in the woods was naturally captivated and carried away into the great, active world that he had not before dreamed of.

Lee Burns, historian of the road in Indiana, has noted that before its coming people had traveled chiefly north and south, along the streams, between Lake Erie and the Ohio River. Afterwards, they traveled more on the east-west axis. The road thus must have had an effect in detaching the people of these states from their early southern connections and allying them with the East instead of with slave-power of the lower Mississippi Valley. As we must always remember, the most important freight that a road carries may be neither household goods, nor live-stock, nor munitions of war—but ideas!

The road served as a line for settlement throughout its length. Taverns, some of their buildings still standing, sprang up along it; towns, also. In fact the towns along the road can still be easily classified as pre-highway, highway, or post-highway. The older and larger ones, such as Indianapolis, Terre Haute and Vandalia have streets that do not coincide with the course of the highway, thus showing that they were laid out before it. A few small towns, chiefly in Illinois, have a street other than their main one coinciding with the road, and center rather upon the railroad that runs a little to the northward, thereby showing that they grew up along the railroad, after the road had become of secondary importance. The great majority of the smaller towns, however, have main streets that coincide with the direction and width of the original road, an indication that they were laid out along it. Even the outlying, more recently developed districts of cities, as at Indianapolis and Terre Haute, show the same centering along the road.

Moreover, even though the road was never finished, its influence upon the modern U. S. 40 is tremendous. From this point of view the question of its completion was of little importance, because its pavement would by now have been wholly superseded in any case. What is much more important is that the road was actually located, so that its broad and straight right-of-way was preserved for the future. From Columbus to Indianapolis and from Indianapolis to Terre Haute and Vandalia, the surveyors seem to have laid out the road on a compass-bearing determined by the preliminary survey. The refinements of modern highway-engineering can do nothing better. From Columbus to Indianapolis this line is about six and one-half degrees south of west. From Indianapolis to Vandalia it is about twenty degrees south of west.

The country is mainly level, but streams are likely to be channeled out fifty feet or more below the general surface. Rarely, except at the stream-crossings, does the road bend. Having crossed the stream and attained the general level of the country again, the road straightens out to its course.

The width of the right-of-way has also been of great importance. It was set originally at eighty feet, and so is wide enough, in a pinch, for a four-lane boulevard. This right-of-way was maintained even through the Dark

Ages. Its existence has certainly helped U. S. 40 to approach close to perfection across the whole breadth of Indiana, and to be a finer highway over a longer distance in that state than in any other. . . .

The story of the National Road in the period of railroad domination is much the same, east and west.

In Ohio the road became an unwanted baby that was left on anyone's doorstep, to be passed on to someone else as soon as possible. In 1831 the state accepted the completed portions from the federal government. In 1854 the state leased its share of the road to private turnpike companies, but business was so poor that the companies soon passed the road back to the state. At last, during the seventies, the state handed the road over to the cities, villages, and counties through which it passed, and it reached the final nadir of being wholly local.

Elsewhere, also, the road was given to the states. In 1849, during the vogue of plank roads, a long stretch on both sides of Indianapolis was consigned to a turnpike company and was surfaced with oak planks. In Wayne County, the easternmost in Indiana, a private toll-road was operated for more than forty years.

As with the original road from Cumberland to Wheeling, this western extension was, in many parts, so well built, and was such a basic route of communication, that it was never allowed wholly to deteriorate.

When railroads were built in this area, across Indiana and Illinois, they paralleled the line of the road, since no shorter or better route was possible. The usual historical sequence is thus reversed. In the United States, when road and railway run side by side, one can usually assume that the railway came first, for in most instances early roads were too crooked or too hilly for railways to follow. But along the National Road across the flat country the opposite holds.

Just before the automobile era the right-of-way was threatened when interurban trolley lines were allowed to build along it. Such lines once extended for many miles on both sides of Columbus. These companies, however, wilted before the competition of buses and private automobiles, and the right-of-way was gradually freed of their tracks. At no place on its whole length is U. S. 40 now so paralleled. . . .

As contrasted even with the never-finished part of the National Road in Illinois to the east of Vandalia, the little stretch of forty-seven miles between

Vandalia and the Mississippi River is something else again. Preliminary surveys were made over this sector, but no clearing was done and no right-of-way established. West of Vandalia the main route of travel was, until quite recently, a little road that had developed for local needs. You can still follow it, if you take U. S. Alternate 40 west from Vandalia to Greenville. It winds charmingly across the slightly rolling prairie country, passing from farmhouse to farmhouse, accommodating itself to the position of woodlots and pastures. It is in full contrast to the bold straightaways to the east. As a result, to accommodate the modern four-lane highway now being built, almost complete relocation has been necessary.

This stretch has remained, even to the present time, one of the poorest sections on the whole transcontinental reach. Although much of it had been relocated as a wholly new highway, even in 1950 other sections remained narrow and poorly paved. In that year travel-services were routing Indianapolis-St. Louis motorists by way of Springfield, Illinois, over Highways 36 and 66.

The continuing poor condition of this part, a direct route between neighboring large cities, is surprising until we consider the political situation. Indianapolis is in Indiana and St. Louis is in Missouri, but most of the road lies within Illinois. This section of Illinois, moreover, is thinly populated, with small towns and rather poor farms. There is, therefore, little pressure within the state for the expenditure of large sums upon U. S. 40. Similar situations have arisen in other places, as where U. S. 91 passes through an uninhabited corner of Arizona connecting Utah and Nevada. Under such conditions the existence of the Public Roads Administration becomes important—to insist upon the expenditure of federal funds for national ends, without regard to the local situation.

In 1919, freshly discharged from the army, I hitchhiked westward from New York, and traveled along the National Old Trail, as it then was known, from Washington, Pennsylvania, to St. Louis. The old S-bridges were still in use. As far as Terre Haute the road was, according to standards of the time, well paved and heavily traveled. But at the Illinois line the bottom dropped out of it. Only an occasional Model-T braved the morass, and the ruts, between towns. The redbud was in bloom and beautiful along the streams, but the road itself was unspeakable. . . .

Both in time and in space, therefore, the National Road is anticlimactic. It started at its height and gradually declined, decade after decade, as the road-bed deteriorated and was not repaired, as constitutional doubts arose—most of all, as canals and railroads were extended. It deteriorated also from east to west so that a great federal undertaking, which the flights of orators had predicted was to reach the Rocky Mountains, merely bogged down in the prairie gumbo.

The doubly melancholy ending should not make us underrate the importance of even this western extension. At its best, in time and in place, it attained greatness. Moreover, it is safe for the future. From Columbus to Indianapolis and from Indianapolis to Vandalia the original route, whether called U. S. 40 or something else, will furnish the main-traveled road until someone disproves the geometrical proposition that a straight line is the shortest distance between two points.

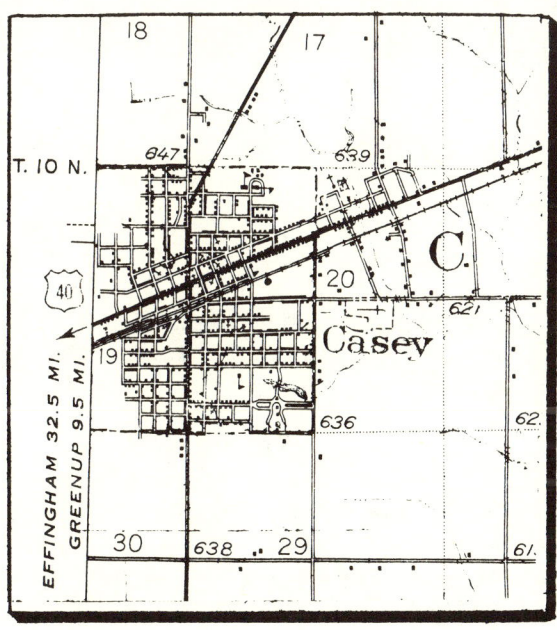

## ㉒ Cambridge, Ohio

🛣 U. S. 40 passes through Cambridge along Wheeling Avenue, a street that like many others in the United States takes its name from the town to which it heads. This picture, looking west, taken at eight on a misty September morning, with the first autumn leaves fallen, displays much of the atmosphere of the smaller American town—even to the informality of the two sitters on the curbstone.

Three eras in architecture are visible. The simple gable and plain lines of the painted brick building immediately across the street bespeak an early date. Actually it was built about 1842, and was intended as a tavern to serve traffic along the National Road. Later it housed a large store known as the "Red Corner." By the enduring tradition of the small town the name still clings even though the building is no longer red. Now, with the old front cut away for display-windows, it houses a "grill," a shoe store, and the office of the power-company.

The heavily corniced building, just to the right, represents another period. Built in 1884, it is of the common style of that time, a period of extensive construction which has left its mark upon nearly all American towns and cities. (See also the picture of Marshall, Illinois.)

At the extreme right, the large five-story building is the Central National Bank. Built in 1906, it is an example of the early skyscraper period. Such baby skyscrapers have been erected in many small towns, in imitation of the larger cities, even though land values scarcely justified and certainly did not necessitate such height.

The Civil War memorial is also a mark of Middle Western towns. This one, like most of them, has little to recommend it artistically, but it at least leaves no doubt that the infantryman ruled the battlefields of that war. Wearing his characteristic cap, he holds the top of the column; a sailor faces us; a curiously paunchy cavalryman with slouch hat and short jacket presents his profile; an artilleryman is on the opposite side. In front of the column sit a woman and a child, allegorically representing: "Knowledge teaching Youth."

Around the octagonal base are chiseled the names of battles—Gettysburg, Atlanta, Winchester, Antietam, Chickamauga, Wilderness, Vicksburg, Shiloh. Except for Chickamauga, these can rate as Union victories. The almost equal division of Ohio troops between eastern and western armies is signalized by the inclusion of four eastern and four western battles.

Ironically, right in front of this War Memorial stands the sign "Wanted Volunteers."

The gigantic elm tree shadowing the square suggests an age well over a century, and is recorded as having been a large tree as early as 1840. It must therefore antedate the arrival of the National Road at Cambridge in 1826, in all probability it was already well established in 1806 when the town was founded. Possibly the street was actually laid out to pass this tree.

The stone milepost at the curb is one of the original posts set along the National Road, and indicated a location 180 miles from Cumberland and 24 miles from Zanesville. It was moved and reset at its present location in 1929, and is no longer of practical value for mileage, being in addition almost illegible.

# ㉓ S-Bridge

🛣️ Across little Fox Creek, just west of New Concord, Ohio, still stands one of the famous S-shaped bridges for which the National Road in Ohio was once famous. Many of these remained in use until well on into the automobile period, although they represent almost the perfect visualization of a modern highway engineer's bad dream. Actually such bridges occur in other states also, for instance in Pennsylvania, but they seem to have been commonest in Ohio.

Although by-passed by the new highway, this particular bridge is still usable, and its S-shape appears as clearly from the marks of the tracks that cross it as from the lines of the parapet. This bridge was built in 1828, and its solid masonry still stands, although saplings are beginning to sprout along the parapet and will soon crack it.

Various folktales are told about the S-bridges. One is that they are really

Z-shaped, and are thus a memorial for Ebenezer Zane. Another inevitable one is that the designer was drunk at the time. One variant of this latter is that the designer was an Englishman and the builder an Irishman. The two met in a taproom, and the Irishman boasted that he could build any bridge an Englishman could design. After a few more drinks the Englishman retired and came forth with the plans for the S-bridge, which the Irishman thereupon constructed.

In 1937 Mr. J. J. Swanson of the Ohio Highway Planning Survey investigated the S-bridges. One of them, near Hendryville, he found to be constructed on such completely irrational lines that he was inclined to favor the persistent Englishman-Irishman story. As far as the other bridges are concerned, his conclusions are wholly rationalistic, and are based upon the fact that the stone arches of the bridges are all so constructed as to cross the stream at a right angle to its course. The reasons for this procedure are obviously economic. A bridge thus built is the shortest one possible, and so uses the least material. It also demands only a simple arch, not a skew arch (angle of arch not at right angles to line of bridge), and a simple arch calls for simple stone-cutting and masonry. After the arch had been so constructed, the wingwalls could be angled off to the line of the highway at both ends. Since even the fastest moving traffic was hardly at more than ten miles an hour, there was no safety-factor involved, and the traveler was scarcely inconvenienced. The bridges of the National Road were a considerable financial problem to the straitened federal finances of the time, and the hard-pressed engineers entrusted with their construction had to save every possible penny.

At this point the new highway has been slightly shifted from the line of the National Road. In the distance, however, it angles back into the old route.

Construction is here so new that the four-lane highway has not actually been opened to traffic, as is evidenced by the road-block in the distance. The rather startling appearance of the large truck proceeding down the wrong lane is therefore fallacious.

The topography here shows the Allegheny Plateau beginning to flatten as it approaches the Central Lowland. The slopes are here gentler, the valleys wider, than in West Virginia and Pennsylvania. Oaks and elms dominate the tree growth.

**24** Highway and Tree

(40) The picture, taken eastward from a point about a mile east of West Jefferson, Ohio, shows U. S. 40 following the line of the old National Road. A heavy rainstorm has just passed over, and the truck is throwing a spray of water.

The highway situation here is both good and bad, and probably represents an attempt to do the best possible in the most economical way with an old situation. The original eighty-foot right-of-way of the National Road apparently represents the width between drainage-ditches, and a four-lane highway has been crowded into this space. This has necessitated the use of an inadequate separation-strip of only four feet, and the situation is therefore dangerous for high-speed traffic. In addition, mail-boxes have been allowed to encroach a little upon the shoulders, but otherwise the design is good. The shoulders are adequate in width and well graveled. The drainage ditch is large and well-constructed of concrete. The backslope has been eased off to prevent erosion, and to approach a natural slope, and is well protected with grass. The pole-lines are back from the ditches.

Most interesting and pleasing is the careful protection of the fine sycamore tree. With the re-grading of the highway this tree would have been sacrificed except for its protection by a stone wall, in itself an attractive bit of masonry. Since native sycamores are fast-growing trees, this one probably does not antedate the construction of the National Road.

The telephone-line, on the right, may be contrasted with that one in the coal-mine picture which displayed eight cross-arms. The more modern construction here consists merely of a single cable which may contain several hundred individual wires.

## 25 Mileposts

(US 40) Where the National Road still follows its original course, the motorist passes many old mileposts still standing. In Pennsylvania these are of cast iron, turned out by local foundries and hauled along with six-horse teams to the proper spot at the roadside. One of these is still in place at the boundary line between Pennsylvania and West Virginia. Although apparently cast to resemble stone pillars and much resembling such pillars when painted, these mileposts consist only of the two sides that face the highway. From this milepost the total length of the National Road can readily be computed as 132 miles. In spite of modern relocations the mileage to Cumberland, as compiled from recent road-maps, is only one mile less than that given here. One may note that the state is Virginia, not West Virginia. This milepost was erected long before the separation of the two parts in 1863.

Stone mileposts occur along the road in Ohio. They are often much defaced, but are now tended by the highway commission. The one shown has been painted white, and the names filled in with black. The prominence of the name Cumberland on all the mileposts helped to make that the common name of the road.

The drivers of the Conestoga wagons, proceeding at their two or three miles an hour, would certainly have been mystified by the modern signpost appearing just to the east of the little town of Lafayette. It must be admitted that the modern automobilist, especially at night, is also sometimes mystified, and has been known to turn up in the wrong lane and go in the wrong direction. This particular signboard, it must be admitted, is about as graphic as possible, and presents a minimum of reading.

Some of the western states have found it necessary to place plain signs upon the highway to indicate the direction in which the traveler is proceeding, thus indicating that the modern American tourist has gone a long way from Daniel Boone and Kit Carson. It is rather a humiliation to have to read: "You are now on U. S. 40, and proceeding west."

## ㉖ Tavern

🛣️ Immediately upon the opening of the National Road taverns sprang up along it. These provided refreshment and lodging for man and beast.

As far as the wagoners were concerned—the "pike-boys," as they were called—most of the refreshment consisted of the fiery product of the local distilleries. Evenings in the bar-rooms were jovial and even boisterous, and not infrequently resulted in fights and gougings. Nevertheless, Thomas Searight, who as an old-timer wrote his history of the National Road in 1894, is highly nostalgic about life of the pike-boys, and did his best to record the complete list of taverns and tavern-keepers.

Most of the taverns were of wood, a fact that may be inferred from the distinguishing name of the one still standing at Lafayette, Ohio. Built in 1837, well within the period of the Classical Revival, the Red Brick Tavern shows the Georgian traditions still strong in American architecture. With its double end-chimneys, it is in fact remarkably like the Mount Washington Tavern shown in the Fort Necessity picture, though that one was built nearly twenty years earlier. Probably an architectural tradition of tavern-building followed west along the road, and there was, as usual, an architectural time-lag from east to west.

When the railroads forced the National Road into obscurity, the taverns generally went out of business. This one followed the same course, as is interestingly shown by the list of Presidents that it has entertained. These include John Quincy Adams, Van Buren, Harrison, Tyler, and Taylor—that is, a group extending in time from the opening of the National Road to 1850. The sixth President to be entertained was Harding, whose dates are subsequent to the opening of the automobile era.

Architecturally beautiful, ivy-grown, shaded by its magnificent sycamore, neatly fenced in white—the Red Brick Tavern suggests the charm of the past. Unfortunately much of the charm is blighted by the enormous and blatantly modern sign which its proprietors have thought necessary to hang on its front, thus throwing the whole effect off-balance. The building has also suffered from being built so close to the original line of the right-of-way. The widening of the old road to a four-lane highway with shoulders has brought the traffic almost into the front door.

The lack of end-windows, except in the gable, indicates a building planned with the expectation that it would be part of a solid row facing a town street. As so often, the town did not grow as solidly as expected. Lafayette, indeed, is one of the many villages that was by-passed by the railroad, and left off the main line of communication. London, a few miles to the southwest and on the railroad, grew up to be the county's chief town. Only after the opening of the automobile era did such villages regain some activity—in this instance, symbolized by the re-opening of the tavern.

The highway, here passing through the village, is another makeshift adaptation—of which Ohio has permitted too many—which crowds four lanes into the original right-of-way. In the distance can be seen the overpass by which U. S. 42 crosses U. S. 40.

The clouds, lacking the sharp edges that are characteristic of the farther West, indicate a humid August midday, with an afternoon thunderstorm beginning to work up.

**㉗ Taylorsville Dam**

**US 40** In western Ohio the red line representing U.S. 40 on the highway-map shows two curious jogs. Near Taylorsville and again near Englewood, the road turns sharply south, west again, and then as sharply north, back to the original line of the National Road. It looks as if the road itself were visibly expressing its disgust at having to turn aside from its straight course. Certainly there seems to have been little accommodation of the highway to the new route, for at both jogs a sharp turn swings the road from its original course and back again, and the pavement is narrow. This part of the road between Springfield and the Indiana line has, in fact, always been a kind of step-child. The original surfacing stopped at Springfield, and for many years the main line of through road-travel passed by the slight southern detour over the locally constructed pike through Dayton and Eaton. Even yet, the unassimilated detours are an evidence that this sector falls well below the average of U.S. 40.

These jogs are caused by two dams of the Miami Conservancy District, erected after the great flood of March, 1913, which inundated the city of Dayton, took 361 lives, and caused damage amounting to more than a billion dollars.

The Taylorsville Dam, here shown from its eastern end, is more than half a mile long, 78 feet high, 415 wide at the base, and broad enough at the top to accommodate U.S. 40 as a two-lane highway.

The picture was taken nearly at high noon, with the August sun blocking out shadow-masses beneath the low arches of the bridge. The Great Miami river, which is here dammed, is allowed to flow through the outlet at normal times, and over the spillway in time of flood. A scale to indicate the height of water shows on the concrete wall.

The grass-covered earth-fill beyond the spillway slopes gently (about one to two and a half) to form a solid barrier, firm as a natural ridge.

Although we are here well within the Central Lowland, and not far from the Indiana line, the forest is still thick, as is shown by the dense woodland forming the background.

## ㉘ In Full Glory

🛣️Seen westward from the railroad overpass half a mile west of the Indiana line, U. S. 40 displays a highly developed modern highway. Except for failing to be a freeway, this section presents no flaws, and since it is designed to serve the countryside as a whole and not merely for through traffic, even that criticism can scarcely be held valid.

The highway here comprises two twelve-foot eastbound lanes (seen directly in foreground), two westbound lanes of equal breadth, and a 48-foot separation-strip. At the edges of the outside lanes are ten-foot, gently sloping shoulders, and four-foot drainage ditches. Beyond the ditches the backslopes rise at a moderate grade to the white post marking the edge of the right-of-way.

The simplicity of line is broken by three crossovers for local traffic, and by the various lanes permitting trucks to leave the highway, stop at the weighing-station, and continue onward.

This highway was constructed in 1947. According to the State Highway Commission of Indiana, if it were to be constructed again, there are no "improvements that would be made over the present design." The highway here completely dominates the country. The ravine just beyond the two billboards has been filled, and the crest of the hill removed.

The blackening of the outside lane from oil-drip is noticeable here as in other pictures. This discoloration is the result of only three years of traffic. The black zone, however, does not center on the lane, for at this point the motorist is just about to drive beneath the underpass, and two signs have warned him. All this apparently produces a psychological need for him to pull over, away from the edge of the overpass.

A weighing-station is often found at state boundaries and is one restriction upon travel. It does not affect the tourist, but raises a problem for the trucker. This is a field in which national regulation might well be applied upon national highways.

On June 13, 1827, the surveyors of the National Road noted that they commenced work "at a stake 2′8″ high on the line dividing the states of Ohio and Indiana, 1 chain and 5 links from a notched beech and 1 chain and 5 links from a notched poplar." These trees stood just eastward of this point. The original line of the road must have run where U. S. 40 is now located. The history of the section, however, is highly complicated because of the existence of the privately constructed Dayton-Eaton turnpike a little distance to the south. The present highway represents the fourth location. One of these, a paved road of 1931, still shows at the right.

The landscape here shows the effects of the ice-sheet that in recent geological times reached down from the north almost to the line of the Ohio River. The low hill in the distance is a deposit of the so-called Wisconsin glaciation.

As always along heavily traveled highways, there is a lamentable scattering of debris, including a long strip of paper in the foreground.

## ㉙ Farm on the National Road

Near Belleville, some twenty miles west of Indianapolis, the old house and newer barns of the Blue Ribbon Farm stand on a little rise to the south.

The house lines up with the highway, and was therefore probably built after the construction of the National Road. The plain lines cry out for an early date, and the building is probably at least a hundred years old.

In this connection Lee Burns comments interestingly. The first houses along the road, he points out, were constructed of natural logs. The next stage involved construction with hewn and squared logs. The third stage was represented by the building of houses from sawed timber or from bricks. Sawmills, however, were far apart, but bricks could be made locally. Brick structure was, therefore, as in this case, very common. Even at this stage there were no architects in Indiana, and as Burns adds, "Perhaps that was an advantage, for the simple lines of the houses built by the settlers after the types of the homes from which many of them had come in Pennsylvania, Maryland, Virginia and the Carolinas are much more pleasing than the pretentious homes of a later period."

All of this is beautifully illustrated by Blue Ribbon Farm. The house is plain, almost stark, but its proportions are fine. The smaller windows of the second floor rest harmoniously over the taller windows below. The general symmetry and the classical motifs of the doorway indicate that the honest builders who designed the façade had known Greek Revival architecture somewhere in the Eastern states.

In certain respects this farmhouse is more primitive than the one in New Jersey already pictured. Noteworthy here are the windowless ends, such as would be expected in a town house, where the builder filled up the whole frontage of his lot. Along with this goes the shallow depth from front to rear, thus enabling the rooms to have the ensured frontage at either side, or in the case of larger rooms to have cross-ventilation. The large chimney at either end bespeaks the cold weather of the Middle West. A sense of good living is also displayed by the location—well back from the highway across a small stream, and on a little rise, for outlook and breeze.

In these days of small families and farm machinery, such a house has become an anachronism. It is close to town, and only half an hour's run by car from a large city. It has thus ceased to be the center of life that it undoubtedly once was. Although the excellent barns and well-painted fences indicate prosperity, the house itself seems to look backward, and shows little evidence of active living. One of the sheltering trees is dying, and the others call for attention.

The large barns indicate a dairy. Of these, the one nearer the highway is of the older type, hip-roofed, probably of the earlier twentieth century. The other, suggestive of a Quonset hut, is an excellent example of modernism in architecture, but still shows the traditional overhang above the loft-door for convenience in hoisting of hay.

The road is here part of the magnificent four-laned highway that U.S. 40 has become nearly all the way across Indiana. The closer lane is probably the older one, but having been re-surfaced with bituminous material, it appears the newer. The broad dividing-strip, although probably mowed early in the season, has now been allowed to grow up charmingly with late-summer flowers.

In the clouds the faint tufts of late morning cumulus, fuzzy at the edges, may be contrasted with the pictures of skies farther west, where the line between white cloud and blue sky is often razor-sharp and paintbrush firm.

**30** Victorian Elegance

**(US 40)** A mile east of Harmony, Indiana, just to the north of the highway, stands a house that may be taken to represent much of mid-Victorian sophistication and comfort. Built in 1872 as the home of a local doctor, it is still owned by his daughter.

Originally it stood well back from the road, which probably occupied little more space than is now filled by the dividing-strip and was at the level of the house. Fine trees stood between house and road. With the widening of the highway and the lowering of its level, the house has been left standing at the very edge of the right-of-way, on a high treeless bank, and a hedge has been grown for some protection. Access to the house must now be by means of the side-road at the right, and the front door has become almost unusable.

The house, almost certainly the work of a trained architect, is an excellent example of mid-Victorian planning, with some French and Italian influence. There is less fussiness than one generally associates with the period, even the wooden struts that support the eaves being comparatively simple. The patterned roof is a typical touch of the time, but this also is not elaborate. The proportions of the façade are admirable. The central archway rises a little higher than the tops of the windows, and the windows are also set a little closer to each other than they are to the doorway. The low windows of the third story admit light and air, without making the house badly proportioned, as it would if the third story rose to full height. The three chimneys bespeak a considerable degree of comfort or even of luxury.

A modern criticism would be that this house is designed for town life, rather than for the country. We must remember, however, that with heating facilities limited to fireplaces, a tight and compact house was much more agreeable during the colder half of the year.

The magnificent tree still shadowing the house gives some evidence of what the effect must have been when the trees in front of the house were still standing.

The highway—recently surfaced, as the pile of gravel at the left indicates—is another example of the splendid construction across Indiana. The grass has been cut back on the lower half of the back-slope.

**31** Roadside Vendor

(40) A great highway like U. S. 40 creates a whole roadside-life along it. There is the world of service-stations and garages, of hotels and motels and motor-courts, of restaurants and cafés and "clubs," of maintenance-stations and police-stations, and of vendors of all kinds, who seek to turn an honest penny in season.

Their products vary with time and place. You can pick up tomatoes in New Jersey; peaches, in Maryland; apples, in Ohio; almost every known variety of fruit, in California. Chenille bedspreads, said to be locally manufactured, hang on lines for sale in the Alleghenies. Maryland offers maple syrup; Utah, honey. Local pottery-ware lines the highway near Zanesville. In season you can pick up, somewhere, practically any kind of garden product.

The crippled melon-vendor here has set up shop just east of the town of Brazil, Indiana—far enough out so that, as the mail-box shows, the rural delivery serves the houses. Someone, a boy doubtless, has hauled the melons there for him in the little wagon, carefully protecting them from being bruised by the sawdust scattered in the bottom of the wagon. He has manufactured a sign—one to face each way—from the sides of a cardboard-carton, and written on it with black crayon, a curious over-correctness prompting him to insert an apostrophe where one is not needed.

But the day has grown hot, and about noon—as the small shadows show—he has dropped off to sleep in his wheel-chair, comfortably in the partial shade.

He sleeps, though the big trucks still go rumbling through. High overhead a thunderhead is building up, but he is safe from rain for two or three hours as yet, and by that time will have had his nap out. He is an old man, as his blotchy face shows. He is sensible to catch a little sleep, even though he may miss a sale or two. Let us wish him happy waking, and pass on—as the photographer did—without disturbing him.

# 32 Benjamin Harrison Era

The town of Marshall, Illinois, has less than three thousand people, but the main business-district is in its way an architectural gem, and might well be preserved as a historical monument. Individually the buildings are about as ugly as can well be imagined, but they were all constructed at about the same time and in the same style, and taken as a whole block, they gain historical significance and even dignity. In a hundred years, if they should be preserved so long, people may be comparing them with the Grande Place in Brussels or some of the crescents in Bath.

The nearest building bears the date 1889; the farthest, 1887. Probably all were constructed within a decade, to judge from similarities of style.

This particular style of architecture of the Early Grover-Cleveland or the Benjamin-Harrison era seems to have no special name. Perhaps it can merely be called Late Victorian. It has been described as a plain brick building with some ornamentation hung on it. The flat roofs are, however, probably the most important feature. These became possible only after the utilization of such roofing-materials as asphalt had been perfected. At the same time cast iron had become cheap enough for common use in construction. The flat roofs seemed to call for some kind of heavy cornice to set them off, and cast iron supplied a safe structural material for the supporting brackets. From the roof, perhaps, the heavy cornices were projected downward to cover the windows. In the picture only the third and fourth store-fronts, in reality the same building, have comparatively simple windows, and even in them the cornice juts out rather heavily. As an example of variety within a prevailing form, it may be noted that no two buildings have the same type of window-cornice, and that in the third and fourth buildings from the left the second-story and third-story windows vary.

As illustrative of a degree of stability seldom associated with the United States, one may notice that the first building was called the Grabenheimer Building in 1889 at the time of its erection, and that the modernized store of the first floor still houses Grabenheimer's.

Not a single tree has been allowed to grow along this block, although during the summer this southward-facing row of buildings must be hideous with heat. This is all the more curious in view of the fact that the residence-district, only a block farther down the street, is sheltered with magnificent shade-trees. Why we Americans should be so dendrophilic in our homes, and so dendrophobic in our business-districts is another subject of study which should be referred to our anthropologists.

Although parking-meters are deemed necessary, the street is almost deserted. This results probably from the time of taking the picture, during the noon-hour.

This is the corner of Archer and Sixth, thus indicating that Marshall, like so many American towns, follows the so-called Philadelphia pattern of street designation, that is, the streets running in one direction are named, the cross-streets are numbered.

## ③ Vandalia

🛣 U. S. 40 passes through Vandalia, Illinois, along Gallatin street, which fittingly bears the name of Jefferson's Secretary of State, sponsor of the original National Road. The final extension of the National Road ended at this spot.

The building is the old state capitol of Illinois, and a highly complicated history lies behind its present handsome appearance. In 1820 the seat of the state government was transfered from Kaskaskia to Vandalia, and in the next few years two state capitol buildings were constructed; one burned, and the other was superseded. A third building, the basis of the present one, was erected in 1836 in a desperate effort by local interests to keep the capital from being moved. As originally constructed, it is described as "Plain, not to say ugly." It had a tiny, "cheesebox-like" cupola, but lacked a portico. The state accepted it in 1837, but almost immediately the General Assembly selected Springfield as the future capital.

After the capital was removed, the building became the property of the county. In 1856 a new cupola was built, and in 1858-9 a thick-columned portico was erected, in the prevailing style of the Greek Revival. In 1899, however, these columns were replaced by the cast iron favored at the end of the century.

In 1918 the state purchased the building to preserve it as a historic monument. Little work was done, however, until the 1930's. It was finally decided "to re-create the interior visual surfaces of the 1836 period, retaining, however, the exterior appearance of 1858." The building as it now appears thus preserves the simple lines created under the influence of Greek Revival in the mid-century, and represents triumph of good sense, good taste and historical research on the part of the Illinois Department of Public Works and Buildings.

Historically, the building is of interest since Abraham Lincoln was a member of the legislature to meet there, and since the act incorporating the city of Chicago was passed within its walls.

The "Madonna of the Trail" statue, at the corner, behind the trash basket, is one of a dozen such placed at intervals along the National Old Trail, to which U.S. 40 has in part succeeded. It represents "a strong featured woman holding a baby in her arms while a small child clings to her skirt as she strides forward on her way." The statues were placed by the Daughters of the American Revolution.

Gallatin Street shows the pleasing patterns of brick pavement. Though now outmoded and too expensive in addition, the old brick surfaces still lend a pleasant contrast to the streets of many Middle Western towns, and in Ohio still extend across a good many miles of the countryside as well. Vandalia is well within the prairie country, but trees grow excellently if planted, as those around the statehouse demonstrate. It has been found necessary to line Gallatin street with parking meters, but the relaxed atmosphere of the small town in summer still is demonstrated by the loose-jointed, overalled citizen, crossing the street in the foreground.

## BOONSLICK, AND MODERN
# St. Louis to Kansas City

The 256-mile trans-Missouri sector of U. S. 40 is marked off at each end by a large city, and also has a political unity in that it traverses a single state and lies wholly within its borders.

Geographically, moreover, it is something of a unit by being the part of the highway that is closely associated with the great inland rivers. It

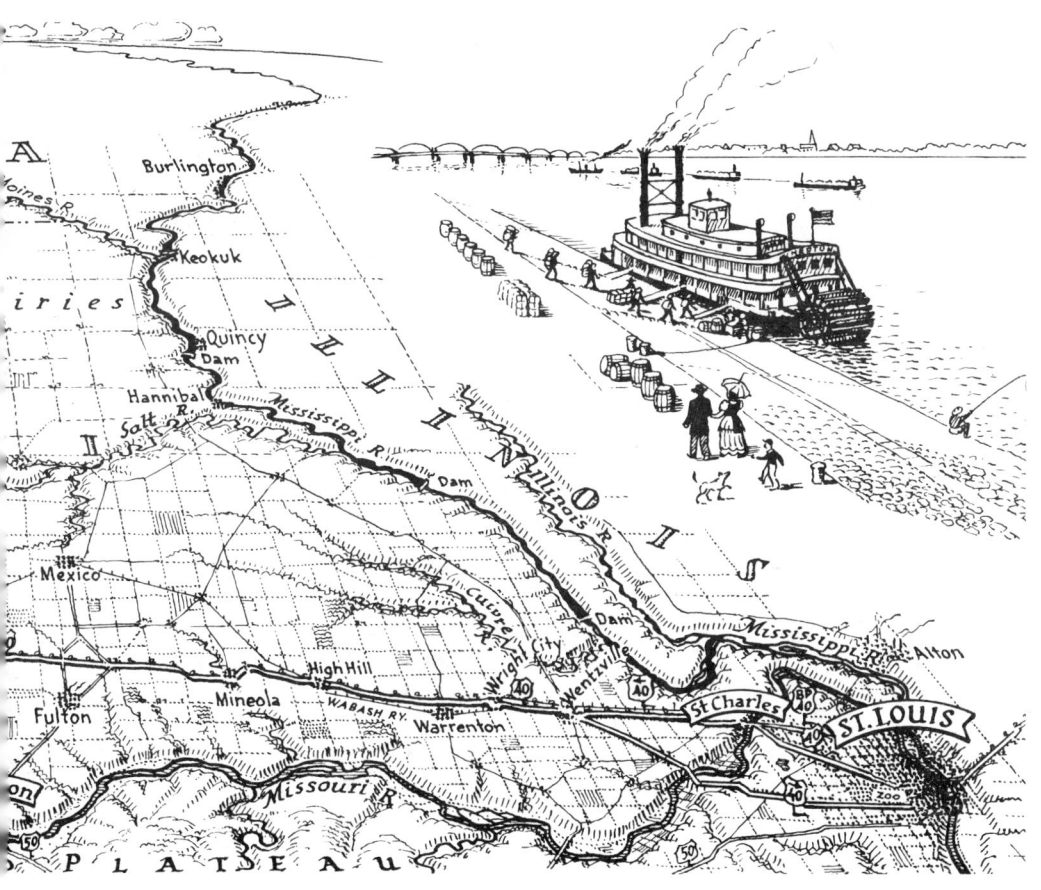

leaves the Mississippi at St. Louis, follows the general line of the Missouri, crosses it twice, and ends at Kansas City, where the highway and the great muddy stream part company.

Topographically the sector is also unified. Although, to speak technically, it passes first across the Dissected Till Plains, and then across the Osage Plains, yet actually it runs all the way in a borderline and transitional area, close to the Ozark Plateau, and generally influenced by the bluff-and-bottom land bordering the Missouri River.

Anyone driving across it will be likely to describe it most easily by the word "pretty." Its eastern half is a land of rolling hills, forested with oak. Here and there it opens up to display fine pastures for cattle. The alluvial bottoms along the river and the smaller streams are rich cornlands. In the western half, after the road has entered the Osage Plains, the country is flatter, and the cornlands stretch out more widely.

Yet, when all is said, nothing stands out. The ordinary tourist, having made the drive, would probably be hard put to tell anything, except that he had twice crossed the Missouri River.

Historically, also, the road is pleasantly interesting, not outstanding. Here, if anywhere, the buffalo-path theory should hold. For, according to the testimony of old inhabitants, Missouri was crisscrossed with well-beaten trails. But no one seems to be sure that U.S. 40 follows such a path.

From St. Louis westward, to accommodate city traffic, the highway splits into two for a distance of forty miles, reuniting at the little town of Wentzville. The recently opened modern road is officially U.S. 40, locally called "New 40," still sometimes mentioned by the State Highway Department as 40 TR (Traffic Relief). By-pass 40 is the old road. As far as St. Charles this latter must follow the route of some trail developed during the French regime. Here, however, an early road would have been of little importance, for in those times the heavy traffic followed the more circuitous river-route.

West of St. Charles, almost to the second Missouri River crossing, the highway follows the line of the old Boonslick Trail. As early as 1790 some kind of road is said to have led westward from St. Charles, but it was certainly of little importance and probably did not run far, for habitually the early French kept to the river.

Traditionally, the establishment of the Boonslick Trail, in spite of the difference in spelling, is attributed to Daniel Boone. Possibly, in his old age, he may have wandered on some hunting trip as far west as central Missouri—there is no evidence that he did not. If so, he probably shot buffalo at one of the salt-licks where they habitually came. What seems more likely, however, is that the lick and the road there take their names from his sons, Nathan and Daniel, who in 1807 with three others went to the lick to manufacture salt. To get there they avoided the cane-brakes and thickets of the bottom lands, and followed the general line of a low ridge some fifteen or twenty miles north of the river. Along the ridge the woods would be more open; the streams, small and easy to ford. The sons of Daniel Boone would have known this elementary bit of woodcraft. U. S. 40 also follows the ridge, and though largely relocated, may in places still keep to the track along which the Boones directed their horses.

Across the western half of the state U. S. 40 is a very modern road. Missourians, indeed, claim that the Santa Fe Trail originally started from Boonville, but even if this somewhat doubtful claim is allowed, we cannot assert that the wagons for Santa Fe ever followed the line of U. S. 40 for more than seven or eight miles. After that point west of Boonville the older route swings off circuitously to the north.

From there on, to Kansas City, the highway runs across open, generally level country, on a direct course. It has many straightaways, one of them twelve miles long. Although these stretches follow the original lines of the land-survey, the highway is not a "section-line road" in the ordinary sense, for there are no right-angled turns and no north-south courses, even for short distances. This section from Boonville to Kansas City best exemplifies what the State Highway Department reports, not without pride: "the major portion of Rte. 40 was built on new location in order to make it as near an air line highway between St. Louis and Kansas City as possible."

The transition of the Boonslick Trail to a wagon-road is not without its interests. As early as 1816, only nine years after the Boones had ridden westward, petitions began to come in from various citizens urging that this or that route west of St. Charles should be established as "the road"— thus, it may be assumed, bringing the westbound wagons past the doors of the petitioners. A law of 1827 established the Boonslick as a state road, but there were still three local roads, each claimed as the genuine one. The matter was decided in 1828, by the choice of the oldest route.

Politically, the situation for U. S. 40 in Missouri is the opposite of that in Illinois. Here the state's two large cities lie at the ends of the route, and there is the strongest kind of pressure for the maintenance of a fine highway. A certain diversion of funds is required for U. S. 50, which also runs between the two cities and passes through the state capital. U. S. 40, however, offers the shorter route. A four-lane highway is now building out from each city, but probably many years are still to elapse before the four lanes meet in the middle of the state and the little towns are all bypassed.

# 34 Mississippi River

U. S. 40, as any transcontinental highway must be, is a great crosser and follower of rivers. Its pavement spans the Delaware and the Susquehanna, and skirts the Potomac without crossing. It crosses the Ohio and most of its major tributaries including the Youghiogheny, Monongahela, and Wabash. It crosses the Mississippi at St. Louis; the Missouri, twice.

It crosses the Kansas and many of its forks; the Colorado, three times; the Green, once; the Humboldt, three times; the Truckee, five times; the Sacramento, once. The aggregate length of its bridges across streams, not counting those bridges which span estuaries, must amount to ten miles at least.

The photograph here shows U. S. 40 crossing the Municipal Bridge over the Mississippi with St. Louis in the background. The river is here about half a mile wide at ordinary water.

The bridge was finished in 1908, and its date is indicated by the solidly built stone piers, for which concrete would have been used a few years later. The pattern of the structural steel also displays the last lingerings of Victorianism in its complicated, at times almost lace-like, design. Nevertheless the three great spans, even if heavy and not highly picturesque, are fully utilitarian. The lower deck accommodates railroad traffic, and a passenger train may be seen half way across. The upper deck carries motor traffic.

The statistics are impressive. In the course of a year the bridge is crossed by approximately twenty-seven thousand locomotives, half a million freight and passenger cars, two million commercial automobiles, and six million private ones.

Beyond the river, stretches the famous St. Louis waterfront, now comparatively deserted. The viaduct carries railroad tracks. The massed buildings of the city's commercial district rise behind the waterfront.

Slightly to the left of the center stands the Civil Courts Building, "the city's most controversial architectural medley," roughly described as rising to the twelfth floor in the style of a modern office-building, there being surmounted by a Greek temple, on which is finally raised an Egyptian pyramid. The striving after height thus satisfied by piling Egypt upon Greece is said to have been inspired by a desire among city officials to outdo the building of the Southwestern Bell Telephone Company, which may be seen slightly to the right. This latter, although a more conventional office-building, has a series of setbacks, which look as if they may have been inspired by some of the more imaginative reproductions of the hanging gardens of Babylon.

Near where the picture was taken, close to the rocks shown in the lower corner—in all the peace that sometimes reigns close to railroad tracks in a great city—an aged Negro was quietly fishing.

## 35 Road-making

*St. Louis to Kansas City* • 153

(US 40) The pick-and-shovel laborers who dug and leveled and surfaced the National Road, as well as the men-driving contractors who bossed them, would be much astonished to watch the process of laying a modern highway.

The photograph shows an up-to-date road-surfacing outfit, capable of laying, on a prepared base, half a mile of concrete in a day. Farthest away, on the as yet unfinished road, a dozen trucks are busily engaged in bringing up sand, gravel, and cement. A tank-truck is supplying water. Two concrete-mixers come next, one of them with its scoop raised high above the skyline, to slide the concrete-mix into the paving-machine immediately in front. After a few feet of clear space is a machine for cutting joints in the slab. Still closer comes finally a leveler which runs back and forth transversely, producing a surface of just the right curve. Only one man is needed to tend each of these latter machines. All the machines run upon the temporary steel tracks at the edge of the slab, which also serve as forms for the concrete until it has hardened. The only actual handworker to be seen is the man in the foreground, who is going over the surface with a straight-edge to get rid of very small variations and to check the smoothness before the arrival of the inspectors. The limit of allowed variation is probably one eighth of an inch in ten feet.

Eventually, after the concrete has hardened and fully cured, the ridge of dirt along the edge of the fill will be smoothed out and a graveled shoulder will be constructed at the edge of the pavement. The backslope, at the right, will probably be planted with grass to check the erosion that has already started.

Only a two-lane slab is being laid, but, with the old road included, a right-of-way sufficient for the eventual four-lane road is being maintained.

The view here is westward about a mile east of the town of Columbia, Missouri. Since there is heavier traffic close to the town, the billboard erecters have chosen this spot to commit another atrocity.

The landscape is rather undistinguished. The forest is mostly small second-growth, with oaks predominating. The smooth skyline shows the general level of the area known topographically as the Dissected Till Plains. Actually the general surface of the country is far from level, because the streams cut down through it. Such dissection, however, does not show in the distant view.

## 36 Boonville

Boonville, Missouri, is described in a pamphlet issued by its Chamber of Commerce as being "where the Boonslick road ended and the Santa Fe trail began." Though the complete accuracy of both these statements may be questioned, the fact remains that Boonville is a highly interesting town.

Thespian Hall, here pictured, may be described in the words of the same pamphlet: "This magnificent structure was built (1855-1857) by the Thespian Society, a frontier dramatic organization dating from the 1830's. It is the oldest surviving theatre building west of the Alleghenies. During the

Civil War it was fortified against attack and was used as a hospital for wounded soldiers and as a military prison. It stands today as a monument to the culture of the frontier and the dramatic art of early Missouri."

It stands also, one is pleased to note, not only as a historical monument, but also as a motion-picture theatre, still at work in spite of its near-century of age, presenting to contemporary citizens the dominating folk-art of the mid-twentieth century. Why it therefore needs to be called the Lyric instead of Thespian Hall, and why such a hideous and unnecessary modern sign has to be plastered up between its two white pillars are questions that we must leave to the anthropologists.

The Greek design of the building is obvious, but the materials would have surprised a classical Greek. The columns are of brick, chiseled into roundness. The superstructure is obviously of wood. The balcony, supported by wrought-iron brackets is more suggestive of French than of Greek influence.

The house beyond the theatre is an almost equally interesting example of mid-nineteenth century Americana. It represents the Gothic revival, which succeeded the Classic Revival and swept across the country in the years following 1860. Gothicism shows in the pointed arch of the porch, in the roof-pinnacles, and in the topheavy chimney. The elaborate cast-iron roof-railing is also Gothic in its suggestion.

The stone marker on the curb commemorates one of the two small Civil War battles fought in the vicinity of the town.

Above the U.S. 40 marker is one of the now rare National Old Trail markers, suggestive in shape and stripes of the national shield.

The telephone-cable in the upper righthand corner not only serves to break up the expanse of sky a little, but also shows a typical feature of an American town. This is a small cable which does not nearly fill up its supporting rings, and consists, probably, of not more than two hundred strands.

The three pedestrians represent apparently mother, daughter, and granddaughter. The photographer's friends declare that the three have been thrown into a panic by observing him, but he maintains the counterclaim that they were disturbed by an approaching car. However it may be, mother seems to have taken charge. Daughter is either passively acquiescent or is expressing her disgust that mother has become so excited. Granddaughter is torn between the two.

## 37 Double Highway

(US 40) In flat country the photographer must always seek some elevated point from which to obtain an adequate view of the highway—a tall building, even an overpass. Here a water-tower supplies the point of vantage.

The direction of view is a little toward the southwest, about a dozen miles east of the Kansas border, and somewhat farther from the outskirts of Kansas City, which is a little outside the field of the picture, to the right.

The parts of U. S. 40 have here been separated by a space of about a hundred yards. The two westbound lanes are in the foreground; the two eastbound lanes, beyond. Although both parts of the road are of modern construction, the eastbound lanes are newer, and demonstrate the way in which the standards of highway construction are constantly rising. The westbound lanes go up and down a little with the natural rolls of the country, but the ruler-straight line of the eastbound lanes shows how cuts and fills have smoothed the country out until the highway proceeds at almost a dead level.

On the whole, this separation of the two parts of a road is not approved by most schools of thought on highway engineering. It is supposed to have a bad effect upon property values in the area left islanded between the two parts of the road. There is also the possibility that some stranger, or even some slightly befuddled native, especially at night, may start in the wrong direction, thinking the road merely a two-lane highway.

The countryside shown here represents the extreme northern edge of that part of the Central Lowland known as the Osage Plains. This area was just too far south to be glaciated, even at the time of the farthest extension of the ice, and it has had a very quiet geological history for a period estimated at a hundred million years. U. S. 40 crosses the Osage Plains in western Missouri, and again between Fort Riley and Salina, in Kansas. The landscape here is slightly rolling, but the general effect is level.

The section here shown is close enough to Kansas City to display certain suburban features. We may note the small power-line in the foreground, the billboard, the building between the two roads (bearing the name of Funhaven), and the house in the foreground. This last, with its newly planted trees in front and its vegetable garden and chicken-run suggests a semi-suburban residence rather than a true farmhouse.

The country is largely in pastureland. Some of it is wooded, but there are also some grain fields. The hedgerow growth is noticeable, and as often many of the trees are elms.

## ⑱ Kansas City

🛣️ This view of Kansas City, taken northward from the tower of the Liberty Memorial, may be permitted to stand for the portrait of the typical American city as traversed by U. S. 40. The highway itself—U. S. Alternate 40, here thoroughly dominated by the city—crosses the railroad tracks via the Oak Street overpass at the extreme right. Closer at hand the Main Street overpass crosses the numerous railroad tracks. Two street-cars appear on Main Street—worthy of note, since streetcars in an American city are already beginning to carry something of an antique or quaint value.

In the foreground stands the Union Railroad Station, completed in 1914, about at the end of the great period in the building of railway terminals. Although architecturally to be described only by that doubtful term "modified Renaissance," and outwardly lacking any trace of modernism, the building is, nevertheless, highly functional. Its obvious T-shape yields good space for lobbies and offices along the head of the T, and easy access to the many lines of track from the sides of the stem, extending over the tracks. The parking spaces and lines of taxis in front of the station may be taken as symbolic of the present-day dependence of railroads upon motors.

In the right-hand lower corner, the cluster of billboards in the vacant lot is also, unfortunately, typical of the American city, and so commonplace as hardly to be imagined otherwise.

Beyond the station stretches off, for some half-dozen blocks, the secondary business district of an American city. The buildings—regional headquarters, warehouses, and so forth—are architecturally utilitarian. They are generally low, and even the highest ones reach up only eight or ten stories, mere flat-topped, almost cubical, boxes of commerce and industry.

Still farther off, all the more prominent because on higher ground, rise the skyscrapers of the central district—the architectural pride of an American city. Farthest to the left, the Kansas City Power and Light Building, of thirty-four stories, towers to 481 feet, Missouri's tallest structure. Next, the tall shaft of the Fidelity Building breaks the horizon line. Just to its right appears a huddled and undistinguished group of lower buildings, marked by the flat-topped style of the early twentieth century. Then the magnificent Southwestern Bell Telephone Building stands alone. At the right are the City Hall and the Jackson County Courthouse, the latter partially concealing the former.

All that has been said about this picture seems perhaps so commonplace that to an American, as with the billboards, nothing else seems quite imaginable. Perhaps the picture will seem less commonplace if we consider what does not show, in contrast to cities of other times and places. An ancient Greek city was conventionally dominated by its "crown of towers," that is, by its military wall. There is no visible mark of militarism in this view of Kansas City. A medieval city had its walls, but its castle and cathedral stood out most prominently. Perhaps we may take, as successors to the castle, the City Hall and Courthouse, especially since the upper three floors of the latter constitute the county jail. But these buildings are not a part of any scheme of fortification, as the castle was, and the analogy is weak. Moreover, no church seems to show in the picture, and this is characteristic of an age and civilization in which religious institutions have generally been dwarfed by secular ones.

### 39 Sign Post

(US 40) Although duplication and multiplication of routes over the same actual roadway, with resulting difficulty as regards signposts, was one of the charges against the old "trails" system, our modern numbered highways have not altogether avoided the same difficulty. Headed west at the corner of Oak and Sixth streets in Kansas City, the tourist finds himself confronted with an array of no less than seven signs on the same post. If at the moment, as is more than likely, he is struggling to guide his car through heavy traffic, he may find himself unable to select the proper number in this array and make the correct turn while there is still a chance to do so.

To add to his difficulties some of the numbers refer to federal and some to state routes. The only order in their arrangement seems to be that the straight-ahead routes come at the top and the right-turn routes below. There seems, indeed, to be a certain progression from small numerals at the top to large ones at the bottom. But this is interrupted by the position of 71. Perhaps 10 is placed lower because it is a state route. Perhaps also the alternates are placed at the bottom. But certainly the harassed stranger, driving his car through Kansas City, does not have time to figure out any such system, and has to let his eyes search wildly up and down, as the horns of aggressive drivers honk behind him.

Fortunately at this corner there are only two choices—go straight ahead, or turn to the right. At other corners, with a three-way choice, the situation is correspondingly more complicated.

Incidentally, a man with a knowledge of the numbering-system of the U. S. highways could tell from this sign that it was close to the center of the country; and he would be certain also that the 10 refers to a state route, since U. S. 10 would be in the far north.

The background shows a conventionally signboard-ridden American street-corner.

Also typically American is the well trained citizen on the opposite curve. One arm on hip, package under the other elbow, she is resignedly and lawfully waiting for that mechanical tyrant the traffic-light to give her permission to cross the street.

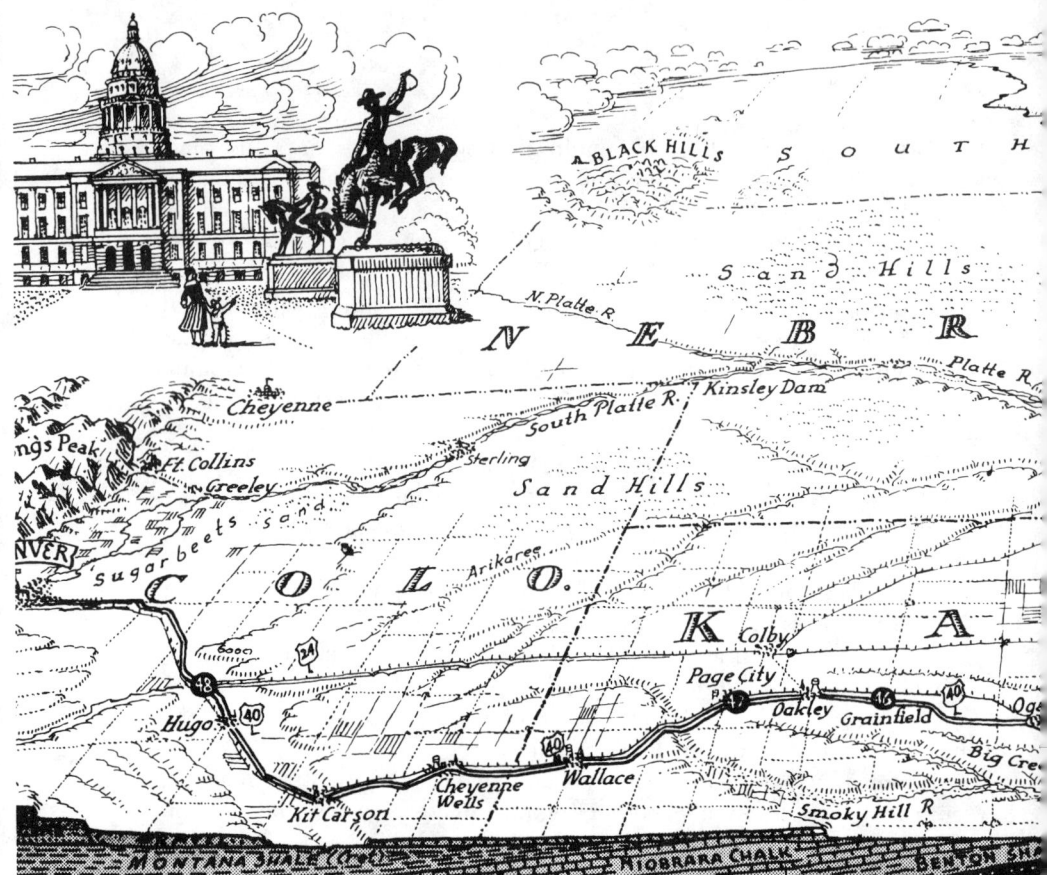

## SMOKY HILL TRAIL

# Kansas City to Denver

S et off primarily by the punctuation-points created by two great cities, the 636-mile stretch of U. S. 40 between Kansas City and Denver possesses also some geographical and topographical unity. At Kansas City the highway leaves not only the state of Missouri but also, approximately speaking, the rolling and wooded lands. Ahead, all the way to Denver, stretches

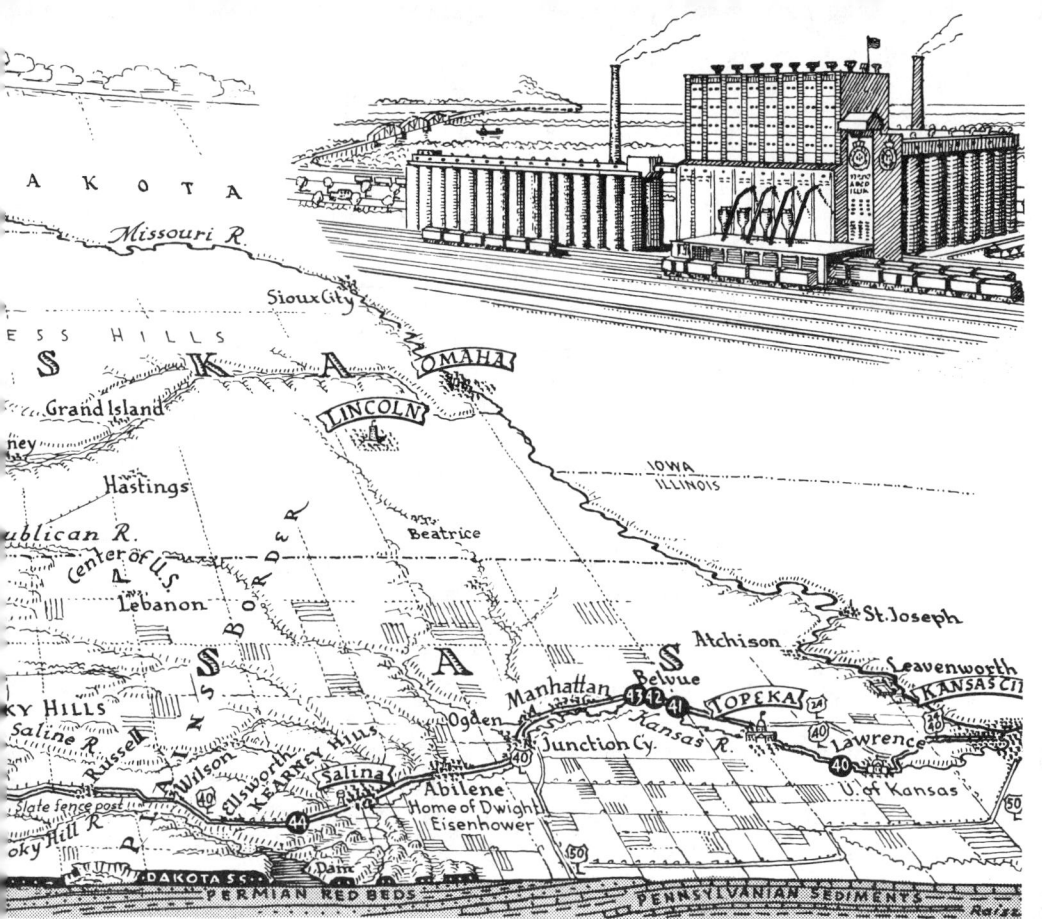

the open country. Physiographers, indeed, distinguish the Osage Plains of eastern Kansas, which are classed as a southern extension of the Central Lowland, from the Great Plains farther to the west, and these again from the slightly rolling Colorado Piedmont. The ordinary motorist, however, observes no sharp boundary and no great difference. In fact, he is likely to describe the drive as "Very dull! All flat!"

Actually, to anyone who knows what he is seeing, this run presents a drama more striking, perhaps, than any to be seen elsewhere along the whole cross section—the gradual drying up of the country, westward.

Mere rainfall statistics tell the story. At the Missouri-Kansas line the annual precipitation is 37 inches. The figure falls off, steadily, westward. In western Kansas it drops to 17; in Colorado, to 14.

Natural vegetation reflects the change. At first, trees are plentiful. Next they shrink back to the stream-courses, and the tall blue-stem grass grows by the roadside. Then trees almost disappear even along the streams and the plains are covered only with short grass—or, in dry years, seem almost devoid of vegetation.

Human adaptation to the land reflects the same change, as the cornfields of eastern Kansas yield to the wheatfields of western Kansas, and they in turn to the cattle-ranges of Colorado.

The fences also tell a story. In the East the usual fence as built by the early settlers was one of wooden rails, and a few of these are still to be seen along U.S. 40 in the hills of Maryland. But even in eastern Kansas there was not enough wood for such lavishness, and farmers planted the Osage-orange hedges, many of which still remain. Farther west, the situation was saved by the invention of barbed wire, but the problem of fence-posts was still acute. In one section of west-central Kansas farmers balked at the expense of buying fence-posts and having them freighted in on the railroad. Instead, they quarried a soft local stone that split easily into rough square columns. West of the town of Wilson such posts predominate along U.S. 40 for about fifty miles.

In its development this sector of the highway falls naturally into two divisions—pre-railroad and post-railroad. The dividing line may be set at the little town of Silver Lake, fifteen miles west of Topeka.

In eastern Kansas the taking up of land preceded the building of railways, and roads grew up from the local settlement-to-settlement tracks, doubtless following much the same evolution already described for eastern Maryland. The characteristic north-south, and east-west straightaways, however, do not appear in the older states. In fact, they are probably found nowhere in the world except in the central and western United States, and in western Canada. To most Americans they are so familiar that they are merely passed off as "section-line roads," and dismissed as uninteresting. Actually their origin goes back very deep into American history.

The pope, we may say, began it. In 1493 Alexander VI proclaimed in a bull the famous Papal Line of Demarcation between Portuguese and Spanish claims; to separate the two, he used a line of longitude. This was a handy way, from the point of view of a pope and his advisory cardinals, sitting in the Vatican, to settle a dispute. Actually, it was highly impractical, for on the face of a continent or an ocean a line of longitude is quite invisible, however prominent it may be on a map.

This method of demarcation, however, has appealed to central authorities ever since that time. The English kings adopted it, and their grants of early colonies often defined them as being from sea to sea between two given lines of latitude.

Congress naturally followed. The boundaries of all states were established, in part at least, by means of latitude and longtitude, and two of them—Colorado and Wyoming—were wholly laid out by this method. Of more importance, the great system of surveying, set up in 1785 to care for the western lands, used the four-square principle of north-south-east-west to determine its six-mile-square townships, its mile-square sections, and its quarter-sections. In the original colonies both roads and land-holdings had been generally adapted to the lay of the land, but from Ohio westward the right-angled system held sway.

As a result, many roads, especially in flat or rolling country, ran according to the cardinal points to avoid cutting through private property, and often went around two sides of the square instead of cutting across the diagonal. Much more serious, even disastrous, was the effect of having farms laid out in a similar manner, without reference to contours. One corner of a man's land might be cut off from the rest by a stream or swamp or ridge. Farmers developed the feeling that fields must be right-angled, and thence came the disastrous habit of running straight furrows parallel to the edges of the field, and encouraging erosion. Five thousand years from now an archeologist may write of us: "For some ritualistic cause, doubtless connected with their religion, they laid out the fields in squares, and the decline of their civilization is to be ascribed to the depletion of their soil thus initiated."

U. S. 40, on the whole, displays rather little section-line structure. The National Road was laid out chiefly through unsurveyed country, but in any case it was required by law to run straight, in a direction somewhat south of west. The Boonslick Trail, also, was put through before any survey-lines were run. Between Boonville and Kansas City, indeed, the highway follows the survey-lines for about half the distance, but it is here a modern location, and wholly lacks the right-angled turns. West of Silver Lake U. S. 40 seldom coincides with section-lines, either because it follows a railroad, or because it passes through mountainous country where topography dominates, or because it has been re-located and straightened. An eighty-mile stretch in eastern Kansas is, therefore, the only important one where the highway still—for the most part, though even here not entirely—follows these artificial lines.

The right-angled turns that the motorist must steer around, and even their pattern on the map, are thus eloquent testimony that the route was in general established after the survey and before the railroad. Actually this later date may be set at 1866, when a railroad was completed to Manhattan; by that time the country, as far as Topeka and somewhat beyond, was well settled.

At Silver Lake, the situation changes sharply. From here on, U. S. 40 in general parallels the Kansas City and Denver line of the Union Pacific Railroad, originally the Kansas Pacific. Occasionally, even farther west, the highway reverts to section-lines, as on a ten-mile straightaway east of Ellsworth, where the railroad follows the river and the highway cuts across the bend. But, on the whole, U. S. 40 in central and western Kansas and in eastern Colorado has few right-angled turns, and not many of its stretches line up on east-west or north-south bearings.

The origin of this part of the route, however, can actually be traced back beyond the railroad, which itself began by following the general line of a road. . . .

"Pike's Peak or Bust!" was the word in 1859. The gold-rush to Colorado was on, and one of the jumping-off places was Leavenworth, about twenty-five miles north and west of Kansas City. The region of the gold-strike lay directly westward. The country in between, though unsettled, was fairly well-known. One obvious thing to do was to take the shortest route by heading straight west. It was all open plains-country, and a horse or even an ox-drawn wagon could go almost anywhere. The chief problem was to find water and grass, particularly at the more arid western end. The solution was to follow the line of a stream, and a good one to choose seemed to be the Kansas River, together with one of its upper branches known as the Smoky Hill Fork.

Goldseekers, therefore, began to press westward along this route. Some of them, greenhorns probably, got into difficulties, lost their way, and ended in very dilapidated condition. A few did not get through at all, but died of thirst or exhaustion in some of the dry stretches. Part of the route got the bad name of the Starvation Trail, and there was even a case of cannibalism. One may suggest, however, that these troubles did not arise because people were on the Smoky Hill Trail, but rather because they were off it—that is, lost.

In 1860 the route was better known, and it continued one of the regular trails by which to reach the now rapidly growing city of Denver. A few years later a stage-route was established over it, and the first stage to Denver *via* the Smoky Hill rolled in on September 23, 1865. Stations, for accommodation of the relay-teams, were established along the route.

Almost immediately there was trouble with the Indians. A stage was plundered and burned in October, though the passengers escaped. A station was attacked in November and burned; two men were killed. The stations were rapidly turned into little fortresses, and some of them flew black flags as a symbol that they would give no quarter to redskins.

Throughout the next few years, the Cheyennes, Arapahoes, and other tribes were constantly interrupting stage-service, attacking stations, and even fighting battles with the troops sent to police the plains. They were particularly outraged because the Smoky Hill country was one of the best hunting-grounds for buffalo. Since this was also the flowering-time of the dime-novel and of the first illustrated magazines, much of the conventional Wild West legend—of galloping horsemen and crackling six-shooters—was developed out of what was happening along the Smoky Hill.

There is, however, rather less of the Wild West than we might expect in an account of an actual journey made in June, 1866, by Bayard Taylor —poet, journalist, and professional globe-trotter. The railroad took him only as far as Topeka. From there on, being used to adventurous travel, he enjoyed his stage-ride, though not the meals of pork-fat and half-baked biscuits. He admired the purple distance of the prairie (and compared it to the English moors); he admired the delicious air (and compared it to that of the African desert); he admired some buff-colored rock (and compared it to Malta); he admired the carpet of wild flowers (and compared it to Palestine); he admired Pikes Peak (and compared it to the Jungfrau). He saw buffalo, antelope, rattlesnakes, wolves, and prairie-dogs, but no Indians, although he met cavalry-patrols on guard and heard tall tales of attacks on stages. He found the few people at the stage-stations living largely in caves and dug-outs, in part for protection against Indians, in part because there was no wood for building houses.

Taylor records a little about the road itself. He crossed the Republican River on a pontoon bridge; the Solomon, by a ferry. The road was deeply muddy in some places; rough, in others. Across the divide between the Smoky Hill and the Big Sandy, in eastern Colorado, he describes it as "a fine, hard natural highway." Taylor mentions his bruises matter-of-factly, obviously taking it for granted that in a western stage-coach the passengers would be tossed around considerably. After four and a half days of continuous staging from Fort Riley, he arrived in Denver.

But the railroad was coming fast. In 1867, only a year after Taylor's trip, it had got as far as Hays. There it lagged for a while, apparently not because of constructional difficulties but until the financiers could mature more ways of making money out of it.

The first train finally got through to Denver on June 24, 1870. The Smoky Hill Trail therefore had a short life, and never really became a made road at all. The railroad did not even follow its line closely across most of western Kansas, but swung well to the north to avoid some rough country.

After the Indian menace was removed, settlement followed the railroad. Little towns grew up about stations, and farmers living in sod-houses broke the prairie soil hopefully for corn and wheat. Inasmuch as there was any town-to-town traffic that did not take the railroad, it generally went by a road that followed along the track. No one did much work on such a road or spent much money on it, for obviously it was secondary to the railroad and of little importance. Occasionally, for no very good reason that you could see, perhaps because the first team happened to be directed in that fashion, such a road shifted from one side of the railroad to the other, over a grade crossing.

After automobiles had begun to venture across plains, even as late as the twenties, the guide-books were likely to describe such a route with the sinister terms "natural prairie road," or "dirt road—bad in wet weather."

Only with federal aid and the establishment of the system of numbered highways did the road in western Kansas and eastern Colorado get away from mud and dust—or gravel, at best—and start to become a paved highway.

Even yet it does not amount to much by our modern standards. The pavement is rather narrow two-lane. The roadbed goes up and down with the little dips and rises of the country, and curves as the railroad curves, instead of striking off boldly on its own as a modern highway should. In compensation, it is generally straight and carries very little traffic. If you slow down for towns, nobody really cares how fast you drive. There are no stop-lights or bad intersections. On the right kind of day—and there are many of them—you can watch the great white clouds piling up, and massing, and ever-changing. Then you inhale the great openness of the country and feel the spirit expand, and—as you keep your foot well down on the throttle—you can see, hour by hour, the East gradually shifting over to be the West.

*Riding the Brahman Steer, Rodeo at Russell, Kan.*
*after photo by Nat'l. Geog. Mag.*

## ⓐ Hayfield

🛣 No portrait of Kansas along the course of U. S. 40—or along any highway—would be authentic without at least one agricultural scene. . . . Here a mechanical baler is circling the field counterclockwise. On the right, in the part of the field already covered, the hay-bales stand or lie, just as they have been dropped from the machine, some of them tipping precariously, as if about to join the others that have assumed the more stable position. At the left the hay still lies on the ground. The field is of alfalfa, but on account of heavy rains much coarse grass and many weeds have grown up among it, giving the hay a coarse appearance.

The machine keeps two men busy—one to drive, one to tend to the bales as they emerge at the rear. The latter squats on a little seat, busy at his work. There are no horses, and indeed one can drive the whole length of U. S. 40 through much of the country's richest agricultural land and see only an occasional horse at work in the fields.

Beyond the hay-field lies a pasture, and in the distance the solid wall of a cornfield. To the right of the road is only rough pasture.

The curving and narrow road here represents the pleasant little backwater of U. S. 40 between Lawrence and Topeka. Both these cities are to the south of the Kansas River, and the chief road of the twenties ran between them, keeping on the same side of the stream. This road was naturally chosen for U. S. 40. Later, however, a fine modern highway was built to the north of the river and incorporated into U. S. 24, which over this stretch thus constitutes a shortcut for through traffic, and is even an easier road for local traffic between the two cities.

U. S. 40 is left as a pleasant little secondary road, winding intimately across the farming country. Such a procedure seems to show the breakdown of the whole system of numbered highways, but since a road is now projected all the way along the south side of the Kansas River, U. S. 40 may at some time be restored to its original position as the chief east-west highway through this area.

The pole-line is an indication of the original importance of this road, for in spite of the completion of the new highway, there was no need to move the poles. This line is unusually large, the poles carrying five cross-arms and two cables.

The countryside is here the level plain a few miles back from the trench cut by the Kansas River to the north, very close to the edge of the farthest advance of the ice. This is the first picture which shows a largely treeless aspect. The region here is far enough west so that tree-growth is becoming scanty, except in the stream-valleys and moister lands. The look of the country to the right of the highway suggests that it was never wooded. A few scrubby trees have sprung up along the line of the fence, where they are protected from grazing cattle. Probably, also, the coming of a farming population has protected the pasture from the prairie fires that used to sweep it, and thus aided the growth of a few trees.

The clouds represent two kinds of weather activity. The cirrus or altostratus, at the left, is produced by some weak disturbance covering a large area. The puffball of cumulus in the center is the result of a local activity, a rising air-current produced by the heat of the sun upon the earth. The time of taking the picture was a little after noon, and by four o'clock a local thunderstorm will be likely to develop from this little cloud.

## ❹ Bit of the Old West

(40) This is the kind of trick picture which this photographer does not ordinarily take, but which seemed justified in this instance as a means of emphasizing the tall grass along the roadside.

The picture was taken, west of St. Marys, Kansas, from a side road. A truck and a car, at the left, mark the line of U. S. 40. The hill represents technically the edge of the Dissected Till Plains, the Kansas River here providing the dissection. The view is across the level alluvial plain of the river, with its meandering course marked by lines of trees. In the distance, what looks like a long low ridge is actually the edge and top of the Osage

Plain, which extends many hundreds of miles to the south, here and there trenched by other broad valleys such as this of the Kansas.

In contrast with the pasture in the preceding picture the countryside here is fairly well wooded. This is because, in this picture, one is looking toward the moist bottom-land in the valley.

The farm in the middle distance has a house close to the highway and almost hidden among the trees. Next comes a modern barn, and farthest back, on the right, an old barn with a typical old-fashioned silo. This country, particularly the alluvial valley of the river, is magnificent farmland, and even the slope of the hill just beyond the grass is growing a good crop of corn.

The "big blue-stem grass," as shown in the foreground of the picture, once covered immense areas in the eastern part of the plains, and is a true bit of the Old West. It afforded magnificent natural pasturage for buffalo, and later for cattle. Like many of the native grasses, however, it did not survive well, under continued grazing. In addition it generally occupied good agricultural lands, and so was plowed under. It is therefore, like much of the other natural vegetation, best seen along rights-of-way, in cemeteries, and in other spots where it has received special protection.

At the center, distinguished by its longer and more feathery heads, are several stalks of equally tall "Indian grass." Flowering golden-rod, and the big-leafed compass-plant with its sunflower-like blossoms just passing into seed, also show along with the grasses.

The picture was taken in mid-afternoon, and there had been heavy rain in the morning. The clouds represent a strato-cumulus cover succeeding the storm-front. The clouds are just beginning to break, however, and there is a distinct suggestion of sunlight toward the southwest, at the left of the picture.

## ㊷ Kansas Corn

🛡 The words of the song, "As corny as Kansas in August," are well demonstrated here, but—to be wholly correct—the picture was taken in September. Except for a single field, everything that is cultivated is in corn.

The pattern of the corn-rows, as shown in the field in the foreground, is so typical of the rural United States that we would feel lost without it. Actually, corn is characteristic of Kansas only in its eastern part; farther west, the country is too dry.

As always in flat country, the photographer must make use of any slight elevation. The vantage-point is here supplied by the overpass on which the highway crosses the railroad. The location is just west of the village of Belvue, which shows among the trees.

The location of the railroad track is in itself of some interest. Here it runs directly east and west for a distance of about five miles, along a section-line. Ordinarily, railroads did not bother to follow section-lines, having the country open before them, and being reinforced by their own land-grant privileges.

The question may also be raised why the highway, in open level country and with no apparent reason, suddenly swings across the railroad. In this particular instance the crossing is not without its reason. A few miles farther on, the railroad runs very close to the river, and the highway would have been crowded out. Having crossed at this point, the highway keeps to the north of the track for many miles. Since this point is west of Silver Lake, we are here in the area where the road generally parallels the railroad.

The highway itself appears at the right. The large treelike mass, seeming to rise from the cornfields, is actually a truck with a load of fodder.

The railroad is here the Kansas City and Denver line of the Union Pacific, which was originally the Kansas Pacific, the first railroad to cross the state. The nearest line of poles carries the telegraph-wires and the wires for the local use of the railroad. It is not a first-class line, as can be seen by the leaning poles and by the fact that several of them have been "stubbed," instead of being reset when they rotted off. The pole-line on the farther side of the railroad carries local power- and telephone-lines.

Note also that the slim fence posts indicate modern steel. Even in this part of Kansas there is hardly enough wood for the local production of fenceposts. Kansas sunflowers are blooming along the fence.

On the skyline at the right a huge silo produces a characteristic protuberance of the Kansas plain.

**43** Roadside Park

**(US 40)** The roadside park has become a definite feature of the American highway. Here the tourists—particularly in family groups—can stop to eat a picnic lunch at one of the tables, let the dog find a tree, and give the children a chance to play about and stretch their legs, while the car cools off in the shade.

Of the states through which U. S. 40 passes, probably Ohio and Kansas have provided the best accommodations of this sort. The one here shown is only a short distance from the overpass at which the preceding picture was taken. In fact, the "Slow to 30 Miles" sign is in view, and the rise of the overpass itself shows in the background.

The highway commission has here apparently taken over a little stretch of ground where in the early days a few cottonwoods were planted. A look at these should warn anyone against the generalization, "There are no big trees in Kansas!" The middle one here shown is about four feet in diameter.

In addition, other trees have been planted—evergreens, Osage-orange, elms, and sycamores. In ten or twenty years a fine grove will be the result. The rainfall is here about thirty-three inches annually, and this is plentiful for many varieties of trees, if they are given a good start and protected against the hazards of fire and grazing animals.

Trees are almost the first necessity of a roadside park, for most tourists pass in the summer and need protection against the sun. As other necessities, however, may be noted the conventional luncheon-table (very functionally designed to provide seats and prevent tipping), the oil-drum for garbage, the water pump at the right, and the "rest rooms" beyond the pump, modestly set among a planting of trees.

To the left of the highway, beyond some newly planted trees, rises the solid wall of corn.

The chief lack here is a fence along the highway to keep children and dogs away from the path of moving cars. For esthetic purposes such a fence could easily be masked by bushes or vines.

## 🄹 Plains Border

🄤 A few miles west of Salina, Kansas, begins a broad belt of somewhat rough and rolling country known to physiographers as the Plains Border. Where U. S. 40 crosses it, in west-central Kansas, this region is about 150 miles broad, lying between the Osage Plains to the east and the High Plains to the west. The region is presumed to have once been a part of the High Plains, and to have been channeled out somewhat by stream action, which is gradually wearing it down to the level of the Osage Plains. Actually, much of it is comparatively level, and seems rougher only by comparison with the extremely flat country east and west of it. Occasionally, however, definite hills break its surface.

The view is here northward toward the rolling country that is known locally as the Kearney Hills. These are much the same as the Smoky Hills, a little farther north, which gave the name first to the Smoky Hill Fork of the Kansas River, and later to the trail that followed along that fork.

The Kearney Hills represent an outcropping of the hard Dakota sandstone, which weathers less rapidly than the rocks adjacent to it and so eventually stands up as a hill. (Two later pictures also show the Dakota sandstone, in much more spectacular formations.)

The scene of this picture is about a hundred miles west of that of the preceding one. Rainfall has fallen off six inches, and is here about twenty-seven inches annually. The upland country in this area has always been treeless, and if buffalo were substituted for cattle, the distant view would be just about what the earliest explorers saw.

Short grass, as seen in the roadside-strip in the foreground, is at this point beginning to predominate over tall grass, and the tall grass fails by far to reach the height shown in a preceding picture, even though this year (1950) was one of exceptionally abundant rainfall. Nevertheless the pasturage is deep and lush.

About sixty head of cattle are visible, all white-faced Herefords, the now predominating breed on the western ranges. The concentration at the left is around a feeding-trough. These are all young steers, being fattened for the market.

The fence shows an interesting alternation of old native wooden posts and modern steel ones.

The highway is here narrow and old-fashioned. It has been laid in two sections with an expansion-joint between the two and a marked transverse expansion-joint also. The curb-like effect at the edge is also old-fashioned, indicating construction of twenty years ago. The purpose of this curb was partly to provide extra strength at the edge of the pavement, and partly to make the highway itself serve for drainage, thus protecting the shoulder against erosion.

The magnificent pole-line is chiefly devoted to long-distance wires, as shown by the double pins and insulators on the lowest crossarm and by the down-brackets on the two upper ones. These arrangements permit crossovers of wires and thus avoid some of the electrical difficulties of long-distance transmission.

**45** Shaded Street

(US 40) Hays, Kansas, began as Fort Hays, frontier military post. It shifted to being Hays City, railhead of the sixties, tough cattle-town of the seventies. Now it is quiet and peaceful, and U.S. 40 runs east and west along its shaded Eighth Street.

This picture may be contrasted with that of Marshall, Illinois (see page 142), but the contrast is really between business-district and residence-district, not between two towns or two regions. As pointed out in connection with Marshall, Americans love trees around their homes, but do not tolerate them near their places of business.

This shaded street in Hays is of particular significance because it is the last as one goes west, or the first as one goes east, along the line of U.S. 40, to display a good growth of trees. Around Hays the annual rainfall is about twenty-three inches, scarcely half of that in Maryland. The country is naturally treeless, but if certain varieties are given a start, by being watered through the first few years, they establish themselves and continue to grow well, as the picture demonstrates. From Hays all the way to the Atlantic Coast residence-districts are likely to show fine shade-trees. On the other hand, from Hays all the way to the Pacific Coast the typical street, even in a residence-district, is almost treeless, and good growth is found only in the more highly developed towns and cities, such as Sacramento, where water is available for irrigation.

The trees growing here are some of the imported drought-resistant elms, along with native poplars and cottonwoods. One problem of trees, which may have something to do with their banishment from business-streets, is shown in the foreground where the roots are heaving the sidewalk.

Other typically American features are the white frame houses, set well back from the sidewalk behind broad front lawns, and the wide parkway for the accommodation of the trees.

A shrewd observer of the American scene might well guess, without being told, that the picture was taken on a Saturday afternoon of September. A few fallen leaves, and the football, indicate the month. The concentration of children, and the family-cars parked in front of the houses, suggest a Saturday. Note also the traditional American custom of going barefoot, and the quite modern American custom of girls in pants.

This picture may also be compared with that of the Baltimore rows, as an evidence that anywhere, in city or in town, children like to get into the picture.

# ㊻ Grainfield

🛣️ Two photographs of Grainfield, Kansas, from different points of view, illustrate a good many local activities.

A picture taken from the highway-level emphasizes things near at hand. The grass by the roadside and the highway signs assume importance. Farther away, in flat country, one sees only those things that themselves reach up above the general height—the telephone- and power-poles, and particularly the four grain-elevators. And above these, as so often in the plains country, towers the magnificent massing of the clouds!

A picture taken from the top of the second elevator minimizes such details as grass and signboards, and presents the grand sweep of the country. Railroad and highway and pole-lines run off eastward in parallel lines into the distance until, like those theoretical lines of perspective, they seem to meet at infinity. Actually all three swing to the left at a point some three miles east of town and proceed toward the next town, six miles distant.

In the foreground the town itself displays much activity. A small freight-train, having left some cars on the main line, is maneuvering upon the siding, with a brakeman nonchalantly swinging from the nearest car. A truck and a sedan wait for the crossing to be clear. Another truck is backed against the loading-platform, where milk-cans are standing.

As might be guessed from the cloud-formations, the two pictures were taken on different days.

## ㊸ High Plains

🛣️ In sheer grandeur and massive scenic effect the most magnificent mountains scarcely exceed the High Plains. Two pictures are again presented, the second taken from the top of the grain-elevator that appears in the first, and giving the empty effect of the country as seen without interruption in the foreground. Both pictures look westward from the tiny town known as Page City.

This region is as flat as any known land-surface. It is the work of streams, which flowing from higher country meandered back and forth depositing their silt. Nevertheless, even the High Plains are by no means as flat as the proverbial billiard-table.

In such country one town is usually visible from the next, and here Winona, five miles distant, shows up on the horizon toward the left.

Page City is only 38 miles west of Grainfield, and the rainfall is about the same. It is still "short grass country," but it looks a little drier, perhaps because of drainage or difference in soil. The scanty and bunchy nature of the grass-cover can be seen in the second picture, at the right.

Agriculturally, the country may be called marginal. The soil is excellent, and in years of good rainfall it yields magnificent crops of wheat. In dry years it rapidly becomes a dust-bowl. Because of the impossibility of predicting dry years and the tremendous damage that can happen to the country because of them, many experts believe that the High Plains should not be plowed at all, but should be reserved for grazing and protected by a permanent sod.

The country here seen is actually not quite as empty as it appears on first glance. On closer inspection one can make out a couple of farms, two metal grain-bins in a field, a small herd of cattle, and a cross-road marked by a line of poles. In the picture showing the elevator there is also a man, if you can find him!

# ㊽ Routes Divide

🛡️ The position is here about a hundred yards west of the point at which U. S. 40 and U. S. 24 divide, or unite.

As originally laid out, U. S. 40 consisted of northern and southern branches from Manhattan, Kansas, to the present point, just west of Limon, Colorado. In 1935 U. S. 40N became a part of U. S. 24. U S. 40S then became U S. 40, maintaining the original designation probably because it followed the railroad, was traditionally the main route, and passed through the more important towns.

Competition for the tourist is hot between the towns along the two routes, and the billboards at the separation are a result. U. S. 24 truthfully declares that it offers the shortest route to Kansas City. The advertisement neglects to state that the distance saved is negligible, amounting to only about twenty miles in a total of well over four hundred to the point at which the routes again unite. The proponents of U. S. 40 advertise even more vaguely "Save Time."

From the point of view of the general public it would be a good idea if private interests were forbidden use of the official numerals, particularly when they are, as in this case, combined with the official shield. Actually, however, only a very naïve tourist would pay any attention to such signs; others would realize that they were serving private interests and were unreliable.

Even aside from the two highway billboards, a disreputable clutter of signs mars the junction of these highways, even though not a house is in sight on the whole landscape.

The topographical region is here the Colorado Piedmont, and the view is typical of those along U. S. 40 in most of eastern Colorado. The skyline is as straight and level as any that can be seen even in the High Plains, but the dip and long roll of the country along the highway distinguish this region from the true plains country.

This is still "short grass country," but rainfall is down to about 14 inches, and desert conditions are being approached. Originally it was probably as much frequented by antelope as by buffalo. Aided by a preserve farther to the south, antelope still are fairly common in this vicinity, and may often be seen grazing close to the highway. They show no concern with the passing traffic, but dash off if a car stops.

Roadside-weeds are prominent in this picture. The disturbance of the earth caused by the building of a highway and the carrying in of seeds with the gravel often produce this growth along a highway, where one may find numerous species of plants that are not native to the country and do not spread back into the fields away from the highway-shoulder.

The fence shows the use of steel posts firmly established. Beyond the signboard at the left is a slat snow-fence to prevent the drifting of snow across the highway.

Ordinarily this country is dry and sundrenched. The season of 1950, however, was a wet one, and growth was correspondingly good. Very heavy rains had fallen during the night prior to the taking of this picture. Low and heavy stratus clouds still linger, and there is a threat of rain in the distance.

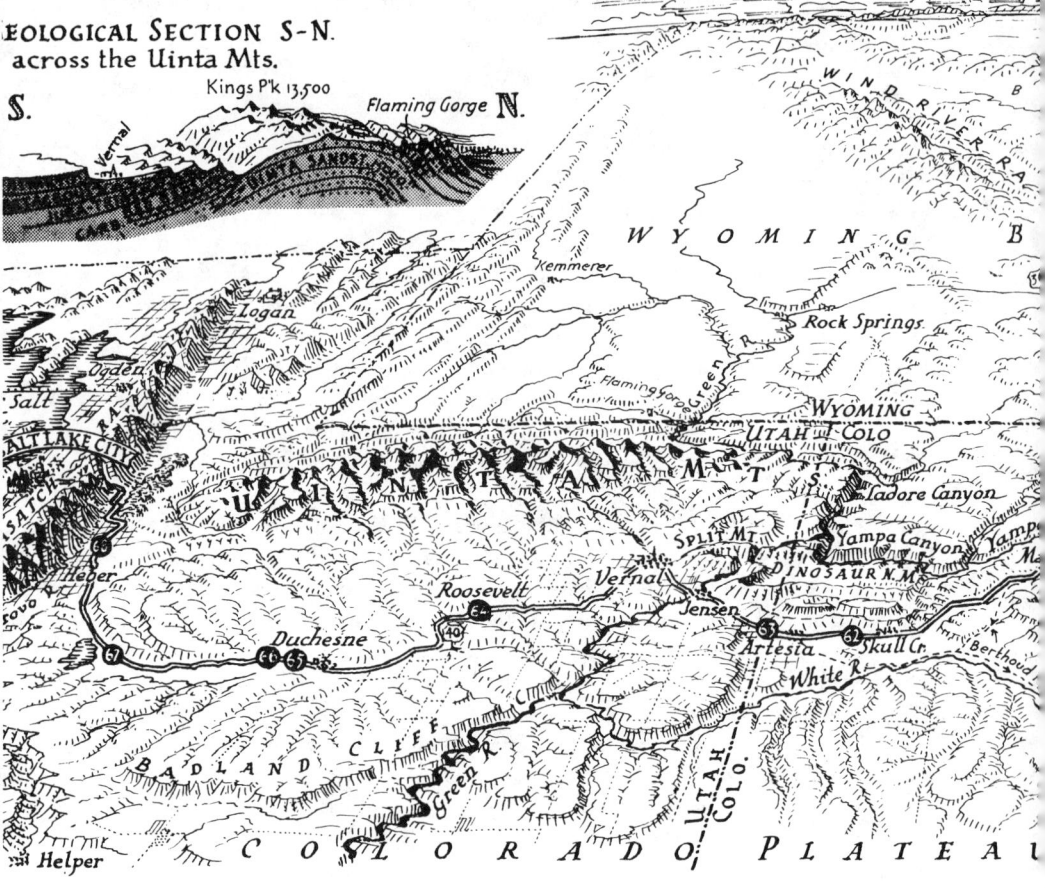

## BERTHOUD'S ROAD
# Denver to Salt Lake City

**B**esides being marked off by a large city at either end, U. S. 40 throughout its 512 miles in western Colorado and eastern Utah is unified by its topography. Just west of Denver it climbs the Front Range of the sharply rising Rockies, and just east of Salt Lake City it drops down from the equally abrupt Wasatch. All the way between, it is essentially a mountain

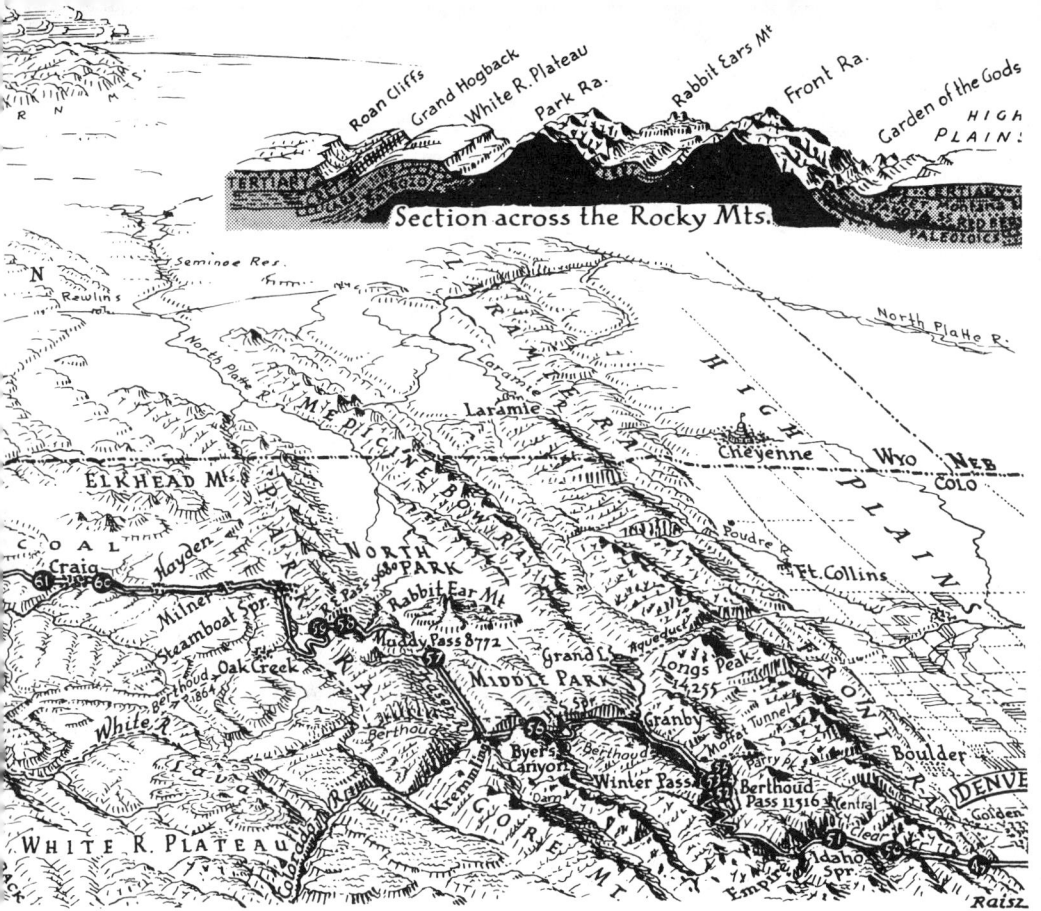

road, and even though it passes for many miles across parts of the Wyoming Basin and of the Colorado Plateau, it is seldom out of sight of some range of the Rockies.

As far as scenery is concerned, this sector is outstanding. At Berthoud Pass the road rises to its highest point, among peaks that are snow-clad during most of the year. Although the Wyoming Basin is comparatively dull, the Colorado Plateau compensates, with its colorful cliffs and canyons of pink-red rocks.

It is all an empty land of far-scattered ranches and tiny towns. It is also a primitive land where you still see log cabins. Much of the country along this part of the highway actually developed within the automobile period. In the now prosperous Yampa Valley the farms were still being homesteaded after 1908. In Utah, Duchesne dates from 1904; Roosevelt, from 1908. Artesia, Colorado, has sprung up within the last decade. These towns, along with all the others between Craig and Heber, a distance of 250 miles, have never heard a locomotive whistle, and probably never will. They are not suffering for a railroad; they are highway towns.

Of the whole cross section of U.S. 40 this sector was the last to be traversed by a road. A few miles in the Wasatch Mountains were opened up by the Donner Party in 1846, and were improved by the Mormons in their great migration of 1847. In 1859 gold was discovered in what was then Kansas Territory (now Colorado), close to the line of the present highway in Clear Creek Canyon, and in that year a road was pushed as far west as the diggings. It dead-ended, however, against the Continental Divide, several hundred miles from any connecting road to Salt Lake City.

Almost immediately the citizens of Denver began to cry for a road across the mountains, westward. This ambition to be on a main transcontinental route of travel was not to be easily satisfied, and for the better part of a century remained a dominating idea that in its local significance may be compared to Russia's urge toward the sea in the world-theater.

In May, 1861, an exploring expedition set out under E. L. Berthoud, who is usually called captain, but preferred to sign himself engineer. (In spite of his obviously French origin, the name of the pass is pronounced with the accent on the first syllable, with an English *th* and the final *d* sounded.) With Berthoud was Jim Bridger, the famous mountain-man.

Berthoud and Bridger did the obvious thing. They headed west, following the canyon, from Empire. They went about nine miles, tried the ridge to the north, gave it up as a bad job. The next day Bridger and some others went off to look for a pass in the vicinity of South Park, but Berthoud again tried closer at hand. With a few companions he went up the north fork of Clear Creek, and then scrambled up the mountainside. At eleven in the morning he reached the ridge, almost exhausted. As he reported: "flowing northwest was a small stream running about fourteen miles through an open, unbroken park."

He then worked along the ridge about four miles east of where he had first surmounted it, and located the lowest spot of a nowhere low watershed. Apparently he assumed—and he was right, though he could not have been certain at first—that the stream flowing northwest went toward the Pacific and did not bend back eastward. He had discovered a high and difficult pass, but he had nevertheless got across the divide.

One may note, in passing, that as far as Berthoud Pass is concerned Senator Benton's buffalo, described on page 17, seem to have failed completely to do the exploratory work that he expected of them. There is no mention of a game-trail across the pass, and no indication that Berthoud or Bridger expected to find such a trail or even looked for one. On the other hand, there was actually a well-worn buffalo-path, mentioned by Frémont, near what is now called Muddy Pass, where U.S 40 again crosses the summit of the Rockies.

Berthoud's party returned to Denver, and reported their discovery, much to the general satisfaction. Another expedition, under the same leaders, was organized to explore clear to Salt Lake City.

Leaving Denver on July 6, the road-finders again surmounted the pass, this time managing to get their horses over. After some thirty miles the explorers came to a considerable river, flowing west. With Bridger's knowledge of geography to aid him, Berthoud must at this point have become certain that he was on the upper waters of what was then called Grand River, since 1921 established officially as part of the Colorado River. From there on, Bridger probably knew at least something about the courses of the larger streams.

The party went on, past some hot sulphur springs, now capitalized into Hot Sulphur Springs. The canyon west of this point was so narrow as to be impracticable for horses, and so the explorers avoided it by keeping to the north, across the hills.

They swung well to the west of the present route to cross the next range of mountains, and then returned to the line of the highway, as they came down to the parklike country along the headwaters of the Yampa River. The scenery, quite rightly, delighted Berthoud, and he wrote:

> Imagine an amphitheatre of long, low ridges, covered with elegant evergreens, meandered by clear brawling brooks, interspersed with

grassy deep green meadows, gemmed with groups of gay purple, red, blue, white and yellow flowers and dotted here and there with small groves, of quaking asps, whose bluish and whitish foliage, ever in motion, gave a magical effect to the landscape.

Through this paradise they followed on, as the highway does, along the course of the Yampa.

Farther west, they swung off toward the south. Nevertheless when they reached Salt Lake City, they had reconnoitered the general course that the highway would later follow. The party returned by a partially different route.

The citizens of Denver were highly pleased. The *Rocky Mountain News* declared on May 27, 1862: "Open this wagon road, and the telegraph and railroad will follow it as sure as the sun rises and sets."

Sure, perhaps, but also slow! There was an attempt to open a road across the pass in 1862, and more work was done in the next year. In 1865 an exploring party, interested in both a stage-road and a railroad, with an escort of 150 men of the Third California Veteran Infantry, left from the Utah end with 22 wagons. They spent nearly four months on the way, but got through. General G. M. Dodge, in a report dated November 1, 1865, declared the route to be practicable, presenting no obstacles that could not be easily overcome, with the exception of Berthoud Pass, "which will require considerable work before it can be used as a stage or emigrant route."

This expedition, apparently, followed the route of Berthoud's return journey. As a result, a line labeled *New Stage Road from Denver to Salt Lake City* was shown on maps published in the next few years—but no stage ever went over it. Conditions were very primitive in 1866, when the never-tiring Bayard Taylor pressed on west of Denver. He looked at the mountains and made comparisons with Italy, Greece, Mont Blanc, Guanajuato, Cashmere, Kashgar, and Kokand. With a horse, but mostly leading it, he crossed Berthoud Pass on June 28, through thigh-deep snow—any trail there was being wholly covered. He got as far as Grand River, opposite the hot springs, but could not cross because of high water.

Two years later, in 1868, Samuel Bowles made a journalistic tour through Colorado, and found the pass still unconquered. He reported: "There was an attempt made a few years ago to build a stage road through the mountains and over into Utah by this route; many thousand dollars were spent upon it, but it was found too big a job." Not until 1874 was a wagon-road actually finished over the pass.

By this time the completion of the transcontinental railroad, with spurs reaching Denver and Salt Lake City, had made a stage-road unnecessary; more than other sectors, this one had to wait for the coming of the automobile. Even local roads were slow in developing, for in this area the Wild West lingered as long as it did anywhere. After the plains Indians had given up the fight, the Utes of the mountains kept their independence, and they ambushed and came close to annihilating three troops of cavalry in 1879, about twenty miles south of the present line of the highway.

Butch Cassidy, one of the last of the rustlers, operated with his Wild Bunch out of Brown's Hole in the northeastern corner of Colorado, and must often have driven cattle across the line of U. S. 40. The Wild Bunch was not broken up until after 1900.

In eastern Utah the transition from rustler to state road commission occurred within a decade. What is now U. S. 40 was designated part of the state highway system, as far as Green River, in 1910.

After World War I this section of the road became part of the Victory Highway. But even when it became U. S. 40 it was still primitive. An improved road had at last been graded over Berthoud Pass in 1922-23. Farther west, however, the guide books still fell back upon the formulae: "dirt and gravel" and "natural prairie road." Special notes had to be made as to where gas and oil were obtainable.

Even yet, this part remains the least traveled and most primitive of any on U. S. 40 There is comparatively little local traffic. The generally undeveloped condition of the highway, especially the steep grades over Berthoud Pass, send the city-to-city trucks around to the north over U. S. 30. The high-speed drivers and even the more timid tourists go that way also.

One should not complain about this or try to change it, but should give thanks. At least there is still left one stretch of several hundred miles, not yet marred by billboards, not yet cluttered with more traffic than it was intended to carry, but offering a sufficiently good pavement and almost continuously varied and even magnificent scenery.

**49** Front Range and Hogback

(US 40) The picture looks north from the so-called hogback, about a dozen miles west of Denver. At the left, the Front Range of the Rockies rises sharply. It is a Sunday about noon, and the highway is dotted with cars, mostly of people coming out from Denver to spend the day in the mountains.

A great deal of geological history shows in this picture. The tremendous uplift of a section of the earth's surface that formed the Rocky Mountains took place along this line. As the mass of land at the left rose thousands of feet, the previously level-lying strata farther to the right were tipped upward until they stood almost on end. The angle of the rock in the foreground shows approximately the position that they assumed. After the uplift—or indeed, during it, since it probably occupied thousands of years —erosion began to work. Soft rocks wore away more rapidly. On the other hand, the Dakota sandstone was extremely resistant to erosion, and was gradually left standing up as the sharp narrow ridge. This hogback runs for many miles along the Front Range, separated from the mountains themselves by the intervening valley. At places, as where the road here circles around its end, it is broken because of local variations in the rock movement.

To the right of the hogback, minor strata of hard rock, showing at the center of the picture, stand up as little reef-like ridges. In the distance on the right, the level expanse of the plains stretches off.

The growth of small trees extending down the ridge is characteristic of such formations. Water soaks into the top and works down along strata, thus enabling trees to grow, whereas on the slope it runs off too quickly. Although there is a thin growth of trees on the mountainside, the valley is largely treeless except where there are plantings. Even the little streamcourse that parallels the highway is almost without growth, doubtless because it contains water for only a short time during the year.

## 50 Two Species

**(US 40)** This picture shows five specimens of *Homo sapiens* observing thirty-six specimens of *Bison bison*.

These bison, more commonly known as buffalo, are kept in Genessee Mountain Park Game Preserve, about twenty miles west of Denver. They consist of males (recognizable by the larger size and more pronounced hump), females, and about a dozen calves.

The specimens of *Homo sapiens* may be further classified as var. *Americanus*, sub-var. *touristicus*. They consist of two males, two females, and a young female. The males may be recognized by their wearing of "pants," a tribal costume originally developed for ease in riding a horse, now grown useless in the automobile period, but retained for its symbolic value. One of these males is engaged in the folk-ritual known as "taking a snapshot." The other is standing with his arms in the position known to anthropologists as "akimbo." All three females are standing with their arms in the same crossed position, a circumstance which may indicate family relationship. Are we justified in assuming that this group consists of two sisters, their husbands, and the daughter of one of the couples?

A neutral observer, such as a bear or elephant, would conclude that *Bison bison* is a dignified and sophisticated species that goes about his business without being at all interested in the antics of *Homo sapiens*, whereas *Homo sapiens*, doubtless because of his simian connections, is highly curious about the doings of other species.

The background shows mountain country, at about nine thousand feet, which would have been natural summer range for buffalo but from which they would have retreated during the winter. The vegetation is native bunch-grass, with a thin growth of ponderosa pine at the higher levels. Even at this altitude these mountains are too dry for luxuriant tree-growth. A three-crossbar telephone-line traverses the shoulder of the slope.

**51** Idaho Springs

(US 40) Like most mining-towns, Idaho Springs, Colorado, straggles along in a canyon beside a stream. The picture was taken from the steep grass- and brush-grown slope to the south of the town. The highway, here reduced to a street, shows at the left.

The general effect of the landscape is that of aridity. At an altitude of 7500 feet, on the eastern slope of the Rockies and cut off by much higher mountains from the moisture-bearing west winds, the town has an annual precipitation of less than sixteen inches yearly. The mountainside shows only a scattered growth of pine—notably thicker, however, on the slopes that face away from the sun, and therefore lose less water by evaporation.

Idaho Springs is now a resort town, but its former greatness shows in the tremendous dumps and the sprawling buildings of the Argo mine. The famous Argo tunnel, which begins at this point, extends four miles through the mountain and emerges at Central City. The mill itself shows the functional design—and the often interesting architectural effect—of a building planned for the treatment of gold. It hugs the slope of the canyon, thus allowing the quartz being treated to descend gradually by gravity.

In January, 1859, the first major gold strike in Colorado was made on Clear Creek at a point to be seen just about at the left of the picture.

The section of the town shown in the foreground, with its small houses of "bungalow" type, generally painted white, is typical of the western small town. Rather curiously the *Colorado Gazetteer* of 1871 describes Idaho Springs in terms which still seem applicable: "the residences...are wooden structures—the lumber from mountain pine—painted white, and neat and tasteful in architecture." The trees growing around the houses are mostly native species such as cottonwoods, resembling the few trees growing naturally by the stream, and reflect the general aridity of the location.

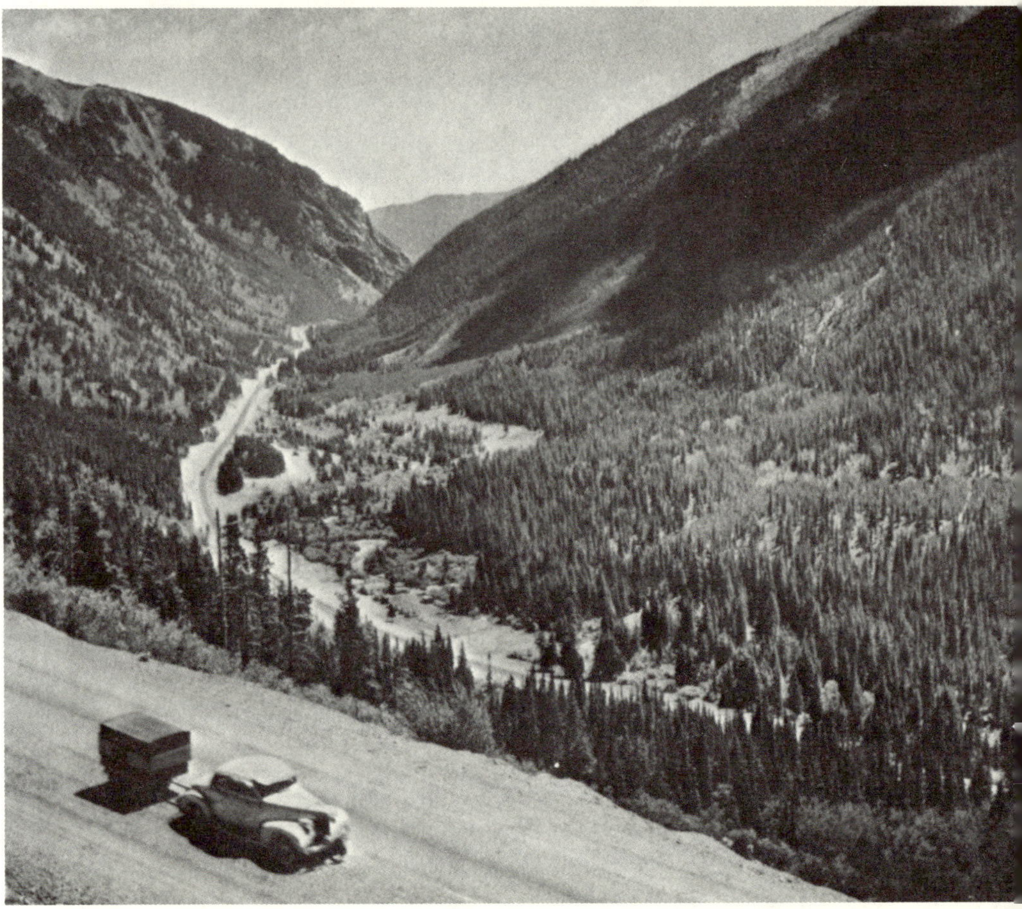

(52) Berthoud Pass—Eastern Approach

**(40)** The picture looks out from the side of the mountain at an elevation of about 10,000 feet. The view is almost directly east, down the valley of Clear Creek. U. S. 40 appears in the foreground, and after making a descending loop of about two miles reappears below.

Although the valley, from this point, appears level, there is actually a considerable slope. Immediately below, the road is at a level of about 9700 feet, but at the lowest point in sight it has dropped 400 feet more.

Western clarity of atmosphere, contrasting with the mistiness and haziness of the Appalachians, is already apparent in the mountains of Colorado. The sharply etched ridge that blocks the end of the valley is actually eight miles distant.

Although the valley lacks the typical U-form, its broad and comparatively level floor suggests that it was carved out by a glacier.

As at Idaho Springs, tree-growth is thicker on the shaded southern side of the valley than on the northern face, which is directly exposed to the sun. The altitude here, however, is so high that the forests are beginning to be influenced as much by alpine conditions as by drought. A very few pines grow along the road at the bottom of the valley. The slopes, however, are clothed almost entirely in spruce, interspersed with patches of aspen, distinguishable by their slightly lighter color. The tops of the ridges are close to timber-line, and tree-growth is scanty.

The aspen-patches usually represent areas where the spruces have been wiped out by fire. The opening immediately in front, which gives a clear field for the picture, is almost certainly such a scar, as a few dead trees to the left help to demonstrate. After the conifers have been burned off, bushes and aspens quickly establish themselves, and may persist for a century or more until the conifers gradually encroach upon them and finally take over—until another fire!

There is thus a kind of natural alternation—for even without man, fires would be set by lightning-strikes. The coming of man, however, always produces many more fires—not to mention logging—so that the scales are tipped against the conifers. Only by the intelligent practice of forestry can the balance be restored. Actually, although the bushes and aspen are valueless for timber, they give feed and shelter to game, especially to deer, and their bright yellow, after the first frost, greatly enhances the beauty of the mountains.

## 53 Berthoud Pass — Eastern Ascent

**US 40** From a mountain-spur 11,500 feet in altitude the view is southwest. Ahead and a little to the left, Red Mountain rises to 12,309 feet. Stanley Mountain (12,516) forming the Continental Divide, rises at the right, its peak not quite in view. The Divide also shows on the skyline in the center around the headwaters of Clear Creek at an altitude of about 12,800. There the divide is about six miles distant in an air-line. Though the time of year is early September, many snowfields are still showing.

The general structure of the Rockies here shows excellently. Notable is the lack of outstanding peaks. The whole long ridge in the background, for instance, does not vary greatly in altitude. According to geological theory, this represents a remnant of the ancient peneplain, which existed at a time when the country in general was elevated to this height but was of comparatively smooth surface. Its present ruggedness has been attained by the wearing down of canyons, so that the actual tops of the mountains remain comparatively level.

The view here extends across several life-zones. As mentioned in connection with the preceding picture, a few pines are growing along the road at the bottom of the valley. These pines represent what is known as the Transition zone. Most of the slope, from the lowest stretch of the road (about 9700) up to timberline (about 11,250) is covered with spruces, and represents what is known as the Canadian zone. Above the irregular limit of tree-growth exists a narrow Hudsonian zone of some scattered shrubs and bushes. Finally there is either bare rock or a low growth of grass, lichens, tiny alpine shrubs, as shown by the little triangle of foreground at the left. This is the Arctic-Alpine zone.

Animal life shows a similar change with altitude. Under primitive conditions mountain-sheep would range in the Arctic-Alpine and down into the Hudsonian. Deer would range upward during the summer through the Canadian, but in the winter would descend to the Transition.

The picture is also interesting for the highway itself. About six miles of it are in view. It enters at the left at about 9700, and climbs about 1300 feet, thus disappearing at about 11,000. The last point to be seen is a mile from the top of the pass, and three hundred feet lower. The picture supplies good evidence that Berthoud is hardly a pass at all.

In his discovery of the route Berthoud went up the valley following the general line of the lowest stretch of the road. Where the road loops back he kept ahead to the right, then ascended the slope somewhat beyond the the shadow-marked side-ridge, and worked back along the main ridge at the upper right hand of the picture.

## 54 Continental Divide

**US 40** The top of Berthoud Pass is a part of the Continental Divide. In the direction of the view, water runs southward into little Hoop Creek, into Clear Creek, into the Platte River, into the Missouri, and into the Mississippi, and finally reaches the Gulf of Mexico. But water falling on this side of the crest runs out toward the Colorado River, and eventually should reach the Gulf of California, unless—as is more likely these days—it is retained behind a dam and runs out eventually from a Los Angeles water-faucet.

In the picture, though it was taken in June, the higher mountains are still deep in snow. The slope shown in the preceding picture is out of sight, immediately below the crest. The opposing snow-covered ridge is actually several miles away, and is the same shown, without snow, in the second picture preceding.

The summit here is at an elevation of 11,315, and is the highest attained by U. S. 40. It is not, however, the only point at which the Continental Divide is crossed, for in this respect U. S. 40 is highly unusual. It returns back to the Atlantic side at Muddy Pass, and then a few miles farther on crosses again to the Pacific side by Rabbit Ears Pass.

The line of dark spruce marching up the mountainside at the left represents almost the highest outpost of the trees. They are enabled to maintain themselves here probably because of the low roll of the ground which protects them from some of the worst of the winds.

The plaque set in the upstanding granite slab records facts of the pass. The wooden signs furnish road directions. Tourist travel is not yet heavy at this time of year; in the late summer there would be a considerable crowd of people gathered about the signs.

The atmospheric conditions, with clouds almost down to the level of the pass, give some indication of the severe weather conditions prevailing at this altitude.

**55** Berthoud Pass – Western Approach

**(40)** Before leaving the summit of Berthoud Pass we look out toward the north. The month, still, is June.

We stand on the Continental Divide, which in its highly sinuous course here runs from east to west. Looping around out of the picture to the right, it then swings north and northwest following the line of the high snow-covered peaks.

Parry Peak (13,345) rises at the right. In the distance is the great mass of mountains within Rocky Mountain National Park, including Longs Peak (14,255).

What appears to be two roads is not the result of double vision but is really all a part of U. S. 40, which in descending from the pass loops back at this point upon itself around a hairpin turn concealed beneath the slope in the foreground. The upper part of the road is sloping toward the viewer, the lower part is really sloping away, although a kind of optical illusion makes it appear at first that both parts of the road are sloping toward the viewer.

As on the other side of the pass, the timber line is approximately at 11,000. The picture shows, however, the extreme irregularity of what is vaguely known by that term. It follows upward in good locations, and drops downward on exposed slopes; so that along this range it presents a scalloped appearance.

The V-shape of this valley would indicate that it is stream-cut rather than glacially shaped. The headwaters of Fraser River flow along the bottom of the valley. This is the northwest-flowing stream that Berthoud saw from the ridge, thus realizing that he had discovered a pass, or at least could so hope.

## 56 Byers Canyon

**40** Just west of Hot Sulphur Springs the railroad, the highway, and the pole-line all crowd into the narrow passageway that the Colorado River has cut through an opposing barrier of hard rock. Such a steep-walled canyon, of which there are many in the West, illustrates the tremendous erosive power of a large stream. Once it has cleared a way for its own course, however, the stream exerts no more influence upon widening the canyon. The atmospheric forces of rain and frost, together with smaller streams, take over the task of widening the canyon into a V-shape. Here, however, the ancient crystalline rocks are so hard that in the millions of years that have elapsed as the river cut downward, the other forces have been able to wear back the sides but slightly. In many places the walls rise almost perpendicularly, and the slopes are so steep that only a scanty growth of pines is possible, as at the right above the freight-train.

A railroad or highway can be taken through such canyon by the simple but expensive process of blasting out a ledge for it. The rock-fragments are then thrown over into the canyon, producing such rockslides as that shown in the foreground. The harder the rock the better, for usually a wall of hard rock, such as that to the left of the highway, will stand up indefinitely by itself, but a soft rock will slide and will have to be held with retaining walls. Even in this canyon there is constant danger of minor slides, because everywhere the material is so steeply piled as to be just at the critical point.

Although such a break in the mountains seems to provide a providential passageway for transportation, originally it scarcely provided a passageway at all. This canyon would have been out of the question for a covered wagon, and probably for a man on horseback. At low water, as in the picture, an active man on foot could probably have worked his way along by following close to the stream, but at high water anyone except a skilled mountaineer would probably have been unable to get through.

In their original exploration Berthoud and Bridger came as far as the sulphur springs just at the upper end of the canyon. Presumably they reconnoitered the canyon, and gave up. Proceeding on horseback, they then crossed these mountains several miles to the north, keeping away from the canyon entirely.

The railroad is the Denver and Salt Lake City line of the Denver and Rio Grande. It is the line which by utilizing the Moffat Tunnel has at last given Denver a direct connection with the West, and it is therefore, along with the highway, the final culmination of the route for which Berthoud and Bridger were exploring.

## 57 Meadow Among Mountains

(US 40) No more typical mountain scene of the West can well be imagined than this one, about half way between Kremmling, Colorado, and Muddy Pass.

To begin, literally, at the bottom, we may note the stream, along which willows, alders, and other small trees grow in such profusion as completely to hide the water itself.

Slightly higher than the stream lies the meadow, in no very ancient time the flood-plain of the stream, and even now, doubtless, occasionally under water. It is still damp throughout such a large part of the year as to prevent any growth but grass. Note, however, the bushes encroaching upon its upper edge at the right.

Above the meadow, beginning at the road and extending to its right, rise the dry bench-lands or foothills. They are thinly covered with sagebrush and bunch-grass, as may be seen in the foreground. Higher, where more moisture is available because of deeper snows, a carpet-like growth of aspens covers the lower slopes of the mountain, and spreads almost over the top of the dome at the right. Still higher, close to the cliffs at the top of the sharp peak, the coniferous growth shows up darkly.

The adaptation of man to the scene is closely adjusted to its varying features. The rancher has placed his house close to the trees along the stream where he has shelter from the bitter winds of winter and a ready water-supply, either from the stream or from a well. He uses the meadow to grow a supply of winter hay for his cattle. At this time, in September, the hay has been cut and collected into two stacks, both carefully fenced to keep cattle from getting at them. Since the harvest is in, he has allowed some cattle to graze in the meadow on the stubble and to pick up the gleanings of the hay. The bench-lands furnish his summer-range for the cattle. In mid-summer they will wander clear up among the aspens and even into the pines.

The highway, too, adjusts itself to the geography. Since the meadow would be boggy at certain times of the year, the original road swung around its edge, keeping a few feet up on the bench. The modern highway similarly swings in a wide semicircle, sacrificing distance for ease of construction, or merely following the old route. Nevertheless, the present road has been much straightened by the cutting off of the points of the little side-ridges. One of these cuts, at the right, is emphasized by a shadow. Directly across the present road to the left of this shadow, the road originally proceeded in a sharp curve.

The view is northwestward. The level of the meadow is about 8000 feet; that of the peak around 11,000. Part of the Rabbit Ears Range, it differs in origins, and correspondingly in form, from the mountains around Berthoud Pass. This range is not the result of erosion from a high peneplain, but comes from volcanic action, so that the mountains are raised up sharply into high individual peaks.

The wooden fence-posts, instead of metal, indicate that we have returned to a country where wood is obtainable as a local product.

In the foreground the fluffy and puffed-up condition of the earth shows where one of the numerous burrowing rodents of the region—probably some kind of gopher—has been at work.

**58** The Rabbit Ears

**(US 40)** Sweeping in a fine curve, U. S. 40 rises on the final ascent to Rabbit Ears Pass, where it crosses the Continental Divide for the third and last time, at an altitude of 9680 feet. The divide is clearly in view. In the foreground and to the right the slope is toward the Atlantic. The opposing slope, visible at the summit, is toward the Pacific.

The Rabbit Ears, a curious rock formation that gives the name to peak and pass, are here visible on the skyline. Since the early days they have been a notable landmark for many miles around. The altitude of the peak is 10,719. An interesting feature is that at a distance the two large formations can be taken for the "ears," but closer at hand—as in this picture—the "ears" are seen to be smaller parts of one formation.

Although the month is June, the grip of winter has still hardly been relaxed on this high country. The sky is gray and bleak. The snow has gone from exposed slopes, and has melted back around clumps of bushes and the trunks of trees from which the heat of the sun is reflected. It is still piled heavily along the highway where the snow-plows have thrown it, and in hollows where it has drifted.

Although the altitude is not excessive for trees, growth is very scanty, probably because at the ridge there is very severe wind-action. Two small lodge-pole pines, recognizable by their smooth light colored bark, stand in the foreground, with a spruce at their right. The lodge-pole, often called a tamarack, is one of the hardiest of the pines, growing at high altitudes, in moist places, and far into the north, springing up rapidly after fires, and having a wide territorial expansion. A few deciduous trees, probably aspens, have not yet put out leaves, and stand as little mist-like patches among the darker conifers.

The power of the wind-action shows in the manner in which the lower branches of the larger pine have been twisted around toward the right, that is, away from the west wind.

Traffic over the pass is low. According to the 1948 count, the latest available for this area, the daily average was only 625 vehicles. This may be contrasted with the 22,688 figure of the six-lane section in Delaware, and the 80,000 of the San Francisco Bay Bridge. The lowest figure for any point on U. S. 40 is apparently the Kansas-Colorado border, where the count in 1948 was only 450, that is, about one half of one per-cent of that on the Bay Bridge.

**59** Stream in Snow

(US 40) The view is eastward, a few miles west from Rabbit Ears Pass. The picture was taken about an hour after the preceding one, and shows the same cold and louring sky, with the clouds not very far above the top of the mountain. U. S. 40, here a narrow and winding and comparatively primitive road, runs along the edge of the trees at the left.

Although the altitude here is 8000, the slopes are gentle and the country so nearly level that the stream is meandering slightly. We are here on the broad, smooth surface of what is known as the South Park Peneplain, one of the surfaces from which the Rockies have been carved. The mountain-wall rising sharply in the distance is—so to speak—a mountain upon a mountain, and represents another cycle of uplift and erosion.

Although this picture was taken on the same June day as the one of Rabbit Ears Pass and a thousand feet lower, the wintry effect is here much more pronounced. Presumably the wind keeps the ridge near the pass comparatively free of snow, but heavy drifts accumulate in this somewhat sheltered hollow along the stream. A few bare patches are beginning to appear at the left, on the slope that faces the noonday sun, but on the opposing slope the snow is still two or three feet deep.

Lack of tree-growth in the broad corridor near the stream may also be explicable in terms of this deep accumulation of snow, although the growth in marginal areas near the tree-line is often hard to explain. Actually the pattern of saplings on the slope at the right—the larger trees higher up and the smaller ones closer to the stream—shows that tree-growth is advancing. In a few years most of this slope will probably be covered with a clump of lodge-pole pines.

The little stream, cutting its way through banks of snow, not only supplies some pretty curves, but also illustrates the erosive action of running water. The little crescent of snow-wall that has recently fallen off in the right foreground would mean, if transferred to the lower Mississippi, the caving away of half a plantation.

No car is visible on the highway, an indication of the small amount of travel across this mountain country, except during the tourist months of July and August.

Note the grayness of the old snow in the foreground, and its channeling into runnels by the melting water.

**60** Oatfield

(40) Drop two thousand feet, shift from June to September, and the deep snows and louring clouds of the high country are transformed into the hot sunlight of a recently-harvested field of oats. The view is eastward, the location a few miles east of Craig, Colorado, in the bottomland along the Yampa River. The highway, as noted in an earlier picture, skirts the edge of the flat, keeping above any danger of flooding.

In this area U. S. 40 really leaves the Rocky Mountains for a considerable distance, although it is seldom out of sight of some of the ranges. This is the somewhat curious interior area forming the so-called Wyoming Basin, which sends an arm southward into this region of Colorado. The basin consists of high rolling country, not really mountainous, and is fairly well represented by the low rounded hills beyond the oatfield. Actually it cuts all the way across between the southern and middle sections of the Rocky Mountains, and by following a circuitous course, a railroad or a highway could avoid the mountains entirely by following through this corridor.

Being at a somewhat lower level and almost wholly surrounded by mountains, the basin is cut off from rain-supply and approaches a desert condition. Much of it, as the name would indicate, drains inward to interior sinks and lakes. The scanty vegetation on these hills shows the low rainfall, which in this particular area is only about fifteen inches a year. This is too low for successful agriculture and the fine crop of oats thus is to be credited partly to an unusually rainy year and partly to the moist character of the low-lying land near the river. The tree and bushes growing close to the highway also indicate some local source of moisture.

The sky with its cloud-formations of cumulus is typical of dry western country in the summer. Although a few showers may fall somewhere, the whole effect is more suggestive of heat and brilliant sunshine than of overcast and rain.

# (61) Sheepherder

US 40 A few miles west of Craig, Colorado, is the proper place to meet Bill Mellos, the sheepherder, with his wagon and horse and two dogs, and his twelve hundred sheep. The sheep, however, are feeding, uphill to the right.

Mr. Mellos and his outfit present an interesting mingling of old and new. He himself, except for the bandanna around his neck, displays nothing conventionally far-western. He wears a battered felt knockabout, not a ten-gallon Stetson. Nevertheless, with the business-like short grip on the bridle-reins, he stands upright, the picture of a vigilant herder, a trifle self-conscious, his glance directed at the far horizon.

The horse, however, scarcely seems to be playing up. At most, he is no agile cow-pony but is more like some comfortable dobbin. Although the ears are alertly up, above the white star in the forehead, there is a general look of sleepiness, and the tight grip on the reins is scarcely needed.

As for the horse-trappings, the rope hackamore in addition to the regular bridle may be noted as a Western touch, and also the high pummel of the saddle, which could still be used for snubbing a rope. One actually hangs at the saddle-bow, but this must be considered a picket-line rather than a lariat. There is no need to rope sheep.

The two dogs refused to have anything to do with the picture. The older black one retired immediately under the wagon. The other, a half-wild four-month-old puppy, made a few threats at the photographer's legs, and then slunk back at his master's command. They are good with sheep, says Mr. Mellos, "Worth more than the sheep."

As for the wagon, all of it above the running-gear is straight out of the Old West. Its canvas cover—ribs showing slightly, weather-proof but light (so that the wagon will not be top-heavy)—goes back to the tradition of the emigrant's covered-wagon, and beyond that to the great Conestogas of the National Road. It is a home as well as a wagon—note the well-blackened stovepipe, the wash-tub, and the worn broom. The mushroom-profile is also traditional and was originally functional, since the old wagon-wheels were higher and the wagon-bed was set low down between them.

The running-gear of the home-on-wheels, however, is modern—taken from an old automobile truck. It shows that the sheepherder is now tied to some passable truck-road. When he has to change location, a truck comes out from headquarters, and tows the wagon. Sheep also are often moved by truck. An attempt to take this wagon across open country might be disastrous, since its center of gravity has been raised dangerously high.

The rolling hills in the background are typical of the Wyoming Basin. Sagebrush is the predominating growth, and to the left of the highway the bushes are luxuriant. The foreground, however, shows a considerable amount of grass between the bushes, and this supplies the feed for the sheep.

Telephone-line, house, and highway, although giving the modern touch, are far from being truly up-to-date. The poles carry only ten strands. The house is primitive, though recent enough to reflect light brilliantly from its new corrugated-iron roof. The rancher has cleared a field in a level spot and, according to Mr. Mellos, is raising "spuds." The herder is disgusted with the whole affair, since it takes some acres of feed away from his sheep. An economic geographer might be equally disgusted, since agriculture in this dry country is a marginal undertaking. In a few years an abandoned field may be developing gullies that will do much more damage than the crops of spuds ever did good.

The highway represents U. S. 40 at close to its lowest ebb—narrow, sharply curving, lacking good shoulders, unfenced, rising at a steep grade. Nevertheless, the daily average traffic (1948) is 900 cars.

# (62) Rock Wall

US 40 At what is known as Rock Wall, in western Colorado, the familiar Dakota sandstone stands for its third portrait. We saw it first as the rolling Kearney Hills in Kansas, grazed across by white-faced cattle. We saw it a second time as the sharply rising hogback along the eastern edge of the Rockies. Here it is again, forming a hogback that is the counterpart, in the direction of its dip, of that one on the other side of the Rockies. Here, as well as there, the strata were rudely lifted from the earth and tilted from their originally horizontal position by the tremendous movement that reared the Rockies high into the air. Between the two hogbacks the Dakota sandstone still occurs in a few places, at much higher altitudes, but in general it has long since been weathered away and eroded by the streams.

As the picture of the other hogback showed the Rockies on one hand and the beginning of the Great Plains on the other, so this one shows the Rockies at the right and the edge of the Colorado Plateau at the left. These are the Middle Rocky Mountains, specifically the Uinta Range. U.S. 40

runs more or less parallel to the Uintas, along their southern edge, for a distance of more than a hundred miles, but is never forced to cross them. They are of interest as being the only important east-west mountain range in the United States.

In this area rainfall is only about ten inches a year, and true desert conditions are approached. The slope in the foreground has a scraggly growth of sagebrush. Two typical desert junipers, traditionally called cedars, stand in the foreground. The clump of trees at the foot of the rock-wall is peculiar, and seems to have been planted, although just why would be hard to say. The higher slopes of the mountains show the darkness that indicates a growth of conifers, but the plateau country is covered only with a scattering of bushes. There is also, noticeable at the right, a fairly good growth of a short grass, now in August ripened and brown.

An isolated ranch is located on the sideroad, about a quarter of a mile off to the right. Rural delivery functions even here, and either Bud Biles or W. S. Stoner (according to the names on the box) has just walked down from the house and is getting the mail. The remarkable thing is that he has walked. Just as in the old days in the West a man used to mount his horse to go across the street, so now he takes his car.

The mail-box is huge—obviously designed to accommodate packages from mail-order houses.

The highway itself here makes one of those curious bends that are so hard to account for. Perhaps originally some property-line made the difference, but there seems at present to be no particular reason why an S-shape should be necessary.

The gully or wash running across the middle of the picture exemplifies one of the present-day western problems. In many places such gullies have developed within fifty years. They are frequently credited to the stripping of vegetation by overgrazing. Another school of thought attributes them to climatic change, and points to evidence that such periods of gullying occurred before the white man had arrived with his cattle and sheep. Whatever the cause, the effect is bad. In what were once good grazing areas sagebrush and juniper are advancing at the expense of grass. Fifty years ago the flat in the middle distance might have been all grass-grown instead of being largely covered with a growth of bushes that are useless for feed.

# (63) State Border

US 40 In the highly federalized United States, since the Civil War at least, state boundaries have had more a sentimental than a practical significance. On a modern highway, as the tourist passes from one state to another, he may note a change of speed-limit, though he will rarely for this reason immediately change the speed at which he is driving. He may discover that the price of gasoline has changed at the line, going up or down a cent or two according to the taxes. Unfortunately for safety's sake, the road-markings may also change, particularly the center-line stripe and its variations to indicate a passing- or non-passing-zone.

At this particular boundary the speed-limit stands at sixty in both states, but the gasoline-tax drops two cents as the motorist enters Utah. Several states along the route of U. S. 40 require only a "reasonable" speed. New Jersey legally permits forty, except where specially posted as forty-five. The sixty permitted by Colorado and Utah is the highest legal speed allowed, and Utah drops this to fifty for night-driving. Gas-tax ranges from $3\frac{1}{2}$ cents in Missouri to $7\frac{1}{2}$ in Colorado.

In some eastern states there is no actual marking of the state line at all, and generally the marking is inconspicuous, but the West lets you know. Here, for instance, we have a farewell to Colorado and a welcome to Utah. The white post at the left is the official survey-marker, but more conspicuous signs have been erected to catch the tourist's eye. As far as the country itself is concerned—barren and wholly without habitation—the shift from one state to the other may be called a purely theoretical transition.

This picture is taken at a point about fifteen miles west of the preceding one. The country differs little, but has perhaps a slightly more arid aspect. To the left of the highway the blackened appearance is the after-effect of a fire that has recently swept across the flat.

At the right and ahead, the precipitous escarpment of the Uinta Mountains rises sharply. At the top of the escarpment, however, the almost level line is conspicuous, as so often throughout the Rockies. The up-lift of the Uinta Mountains is one of the greatest known in geological history, being estimated at 30,000 feet, now worn down until the highest peaks are under 13,000.

A hog-back extends across the left of the picture, and the highway swings in that direction and rises along a fill to cross through a cut in the crest. Actually, by going a little around, the highway could have avoided crossing the hog-back and encircled its end, but the shortening in distance was apparently held to offset the fill and grade.

Visible again is the line of small trees growing along the crest of the hogback in contrast to the barren slopes. The clouds represent a mid-afternoon build-up of cumulus over the mountains. The "anvil" shape, indicating an active thunderstorm, is showing. Rain is undoubtedly falling somewhere on the range.

The flowers blooming rather pathetically in the foreground are roadside weeds that will probably survive for a few seasons until the disturbance of the soil that allowed them to take root is no longer effective and conditions have reverted to normal.

# 64 Roosevelt, Utah

**US 40** The main street of Roosevelt presents the pattern of the far-western small town, with certain adaptations because of its complete dependence upon the highway.

Founded in 1908 and named for the out-going president, Roosevelt was originally a trading-center for scantily inhabited ranching-country. It now subsists largely from highway-income, being the best place between Vernal, 27 miles to the east, and Heber, 99 miles to the west, at which to find a decent lodging for the night, or eat a meal, or buy a new tire, or have a little garage-work done, or amuse yourself by an evening at the movies.

Roosevelt is interesting in that it is wholly a highway-town, having no means of reaching the outside world except over U.S. 40. The closest railhead is fifty miles to the south, and there is no connecting road. The nearest practical contact with the railroad is at Heber, 99 miles away. The citizens of Roosevelt, however, are not at all concerned with their distance from a railroad. They are wholly adjusted to highway-life, and are getting along excellently.

The width of the street, 69 feet between curbs, is typical of the west, and may be contrasted with the narrow streets of Ellicott City and Frederick. Some say that western streets were made wide so that ox-wagons and four-horse stages would be able to turn around. More likely the width may be credited to the boom spirit which saw in every village a future metropolis. Certainly in an unsettled and unforested country there was little to work against wide streets.

A future historian, deprived of other evidence, could reconstruct much of the life of this town from the signs displayed along both sides of the block. Two hotels, two cafés, and a grill indicate its preoccupation with the traveler. Its citizens are also able to buy shoes, drugs, hardware, appliances, and tires. They keep in touch with the dominant forms of American folk-art by patronizing two motion-picture houses, and by listening to radios. Both the height of the bent aerial on the top of the hotel at the right, and the absence of television aerials suggest the distance from large centers.

The architecture—low, flat-roofed, undistinguished—offers little that is outstanding, but is fully characteristic of its time and place. The hotel is carrying on, though in a modest way, the old false-front tradition of the early cow-town. Although no building in the picture can be as much as fifty years old, the general effect is somewhat dilapidated and run down. Like most towns in marginal farming country, Roosevelt has never had money on which to splurge. The showy and new electric street-lights are about the only sign of extravagance.

Roosevelt is a Mormon town, and the surrounding country was settled largely by Mormons who moved to this isolated area to escape the federal anti-polygamy laws. The picture, however, shows almost nothing that is typical of the Mormons. Their love of trees is very imperfectly demonstrated by the single tall cottonwood. Their love of running water along the streets is barely suggested by the deep gutter at the right, although there is actually a running stream beside the cross-street.

The picture was taken about eight o'clock on an August morning, before any clouds had yet had a chance to build up. Through the clear air, at the end of the street, appears one of the long level lines characteristic of the red sandstone plateaus of the region. The gravel scattered upon the pavement, both right and left, indicates that the highway is here only a narrow ribbon through primitive country, and that dirt or gravel roads and streets lead into it from both directions.

## ⑥⑤ Plateau Country

🛡40 West of Duchesne, Utah, U. S. 40 passes through characteristic scenery of the Colorado Plateau. This vast topographical region, here seen along its northern fringe, includes southwestern Colorado and southeastern Utah, and extends well south into New Mexico and Arizona. Not only is it a plateau as a whole, being generally a mile above sea-level, but also it includes many smaller plateaus, such as those seen in the background of the picture.

Generally, in local usage, whole regions are called plateaus, thus preserving the geologists' term. Smaller formations such as those in the picture, more commonly take the Spanish word for table, and are called mesas.

Still smaller ones, where the idea of height is more pronounced, preserve the name introduced by the early French trappers, and are called buttes.

These flat-topped formations, large and small, result from the horizontal beds, also evident in the picture, in which the rocks of the area are laid down. Such a structure breaks off sharply along the edge, thus forming cliffs, but is preserved on the level stretches of its surface. The resulting scenery is often spectacular.

The whole area is a desert, and rainfall in the part represented by the picture is about seven inches a year. The slopes below the cliffs are almost devoid of vegetation, but on the level top a dark spotting of pinyon and juniper trees may be seen.

The valley is more moist, being along the course of Strawberry Creek. Cottonwoods and sycamores grow along the stream, and there is even— with the aid of irrigation—some farming.

Well placed under the shade of the tree at the right, stands a genuine log cabin. The building closer to the road is also a log structure, but with a gable-end of sawed timber. In contrast with the modern reconstruction seen at Fort Necessity, these two are genuine products. The one under the tree actually has a dirt roof. Log cabins are still fairly common along the highway in this region, which was settled without the benefit of a railroad, and even before any automobile communication was established, but less than fifty years ago, so that the original cabins still stand.

Although this area has been considerably affected by the building of a paved highway, it also maintains much of its primitive character. Farms are often small, and like this one represent living at a bare subsistence-level. Note, for instance, the fence posts cut from native trees, and not even leveled off to the same height. Note also the primitive track that runs along between fence and road on the right. This is a track for a team of horses, pulling a wagon or some farm-machinery, which are preferably driven along such a soft track as this, and kept away from the hard pavement.

There is a fine growth of roadside-weeds on the slope of the highway. The higher and clumpy bushes between the horsetrack and the fence indicate the natural vegetation of this flat along the stream.

The tremendous massed cumulus clouds are as characteristic and spectacular scenery of the region as are the flat-topped mesas. At least, they are characteristic of the summer and the afternoon.

## 66 Target of Opportunity

(US 40) A few miles west of the log cabins, three horses—artistically arranging themselves white-black-white—offered a target of opportunity.

The horses themselves, heavy-bodied and fat-rumped, seem strikingly different from the wild cayuses that one expects to encounter on the open range. Nevertheless, they were skittish. When the photographer jumped from his car to get the picture, they immediately took off, and headed across the highway. The approaching truck-driver tramped down hard; brakes squealed; the photographer momentarily envisioned a smashed truck and mangled driver. But the truck squeezed by first, and the photographer had sufficient presence of mind to expose a film at the right moment.

Aside from the circumstances of its taking, there is really very little to be said about the picture. Everyone seems, however, to like it, and it certainly shows a section of U.S. 40. It is therefore included without further apologies.

As far as the background is concerned, the almost level mesa-tops with their sharp falling off at the ends show again the usual skyline of the Colorado Plateau. The nearly level expanse of valley presents nothing but a growth of pinyon and juniper. In fact, this area presents the most extensive such forest in the United States. The stippled effect shown on the most prominent hillside is a common one.

Instead of the piled-up cumulus, which may be called the ordinary cloud of the country in summer, we here have cirrus and high stratus clouds, marking the approach of a weak frontal storm which is sweeping in from the west across the Wasatch Mountains.

This scene displays U.S. 40 at its minimal point in Utah, with a width of only 20 feet and a daily average of cars amounting (1949) to only 775. Note that the road is completely open, with no railroad, billboard, building, fence, or pole anywhere in sight.

### 67 Beaver Dams

**(US 40)** Where the highway ascends the eastern slope of the Wasatch Mountains near Soldier Springs, Utah, the beaver dams supply the chief interest.

Some years in the past a pair of beavers, under the protection of game-laws, decided to set up housekeeping at this point. At that time the aspen forest must have grown clear down to the stream.

Getting to work with the proverbial energy of beavers, this pair erected the lower and smaller dam, and lived happily in a house that they constructed in the pond. They felled aspens by nibbling around the base. They then cut the branches into convenient lengths, dragged them to the pool, and lived upon the bark. The aspen-trunks cut at this time, by beavers living in the lower pond, have now disappeared by decay.

Protected by man and supplied with ample and ready food, the beavers multiplied rapidly, and before long the enlarged family built the larger dam, flooded back the much larger pool, and were ready for further operations. Each night, going to the nearest point at which aspen was available, the beavers cut their trees and obtained their food. Gradually, however, in the course of years, they were forced to cut farther and farther from the edge of the water. They must now go well up the hillside, and the recently cut aspen-trunks still lie there on the ground, as yet undecayed. At the present time, although apparently they are still prosperous, the beavers have possibly reached a critical point, where food-supply is so far from the water that they can reach it only at considerable risk. Such enemies as bobcat and coyote are probably beginning to harry them. The beavers must now perish, or try to build another dam higher upstream, or else go through the risky process of migrating blindly in search of some other stream. A civilization is about to fall.

The picture illustrates also the valuable work performed by the beavers in water-control and the prevention of erosion. Small meadows are gradually forming behind the dams. Even before the meadows are formed, the impounded water leaks out slowly during the dry season, thus maintaining the flow of the streams, and the half-empty dams prevent the downward rush of the sudden flood waters.

In the picture can also be seen the typically Western contrast between the thickly-wooded north slope, and the sage-brush-covered south slope, exposed to the sun, which forms the foreground of the picture.

The wire fence is of local juniper wood, the small posts of which require to be set thickly.

**68** Valley in the Wasatch

**(40)** From a hillside, looking down upon U. S. 40 as it skirts Heber Valley, about three miles north of the town of Heber, Utah, the observer looks somewhat west of south across the lush valley itself and toward the high peaks of the Wasatch Range. The valley is somewhat more than a mile above sea-level. Provo Peak, to the left, twenty-five miles distant, rises to 11,054. Mount Timpanogos, to the right, reaches a height of 12,008, and even in August shows some snow-banks. Through the gap that lies between the two peaks the Provo River flows out from the valley. U. S. 189 and the Heber branch of the railroad also use this gap to pass the mountain-barrier, but U. S. 40 swings off to the north to cross the range and drop down to Salt Lake City.

As far as drainage is concerned, this is already a part of the Great Basin, for the waters of the Provo River never reach the ocean. Topographically, however, the Wasatch Range represents—and rather magnificently—the last western rampart of the Rockies.

In Heber Valley, after the long desert stretches of the Colorado Plateau, the traveler finds himself at last in a green countryside. Although the rainfall is only sixteen inches a year, the valley is well watered from the river, fed throughout the year by the mountain snows. On the level valley-floor the stream splits into numerous channels, and along them the water-loving trees grow luxuriously.

On this side of the trees stretch rich meadows, from which loom the mounds of many haystacks. Still closer lie the rectangles of two barley fields, already white for harvest.

Heber was an early Mormon settlement, and the farm just at the base of the hill displays much of Mormon solidity. The house, built close against the hill for protection against the north wind, is of brick. The barns, though not large, are well built and painted. The sun reflects brilliantly from the corrugated-iron roof of a shed. Where the Mormons settled, they settled with the expectation of staying for a long time—and they have generally stayed. Unseen, though it might be suspected because of the luxurious tree-growth on the hillside, is that other evidence of Mormon occupation, the irrigation ditch that runs along between the point of observation and the house.

The time is early afternoon. In addition to a wisp of high cirrus, cumulus masses are just beginning to form, but they have a long way to go before they will produce a shower.

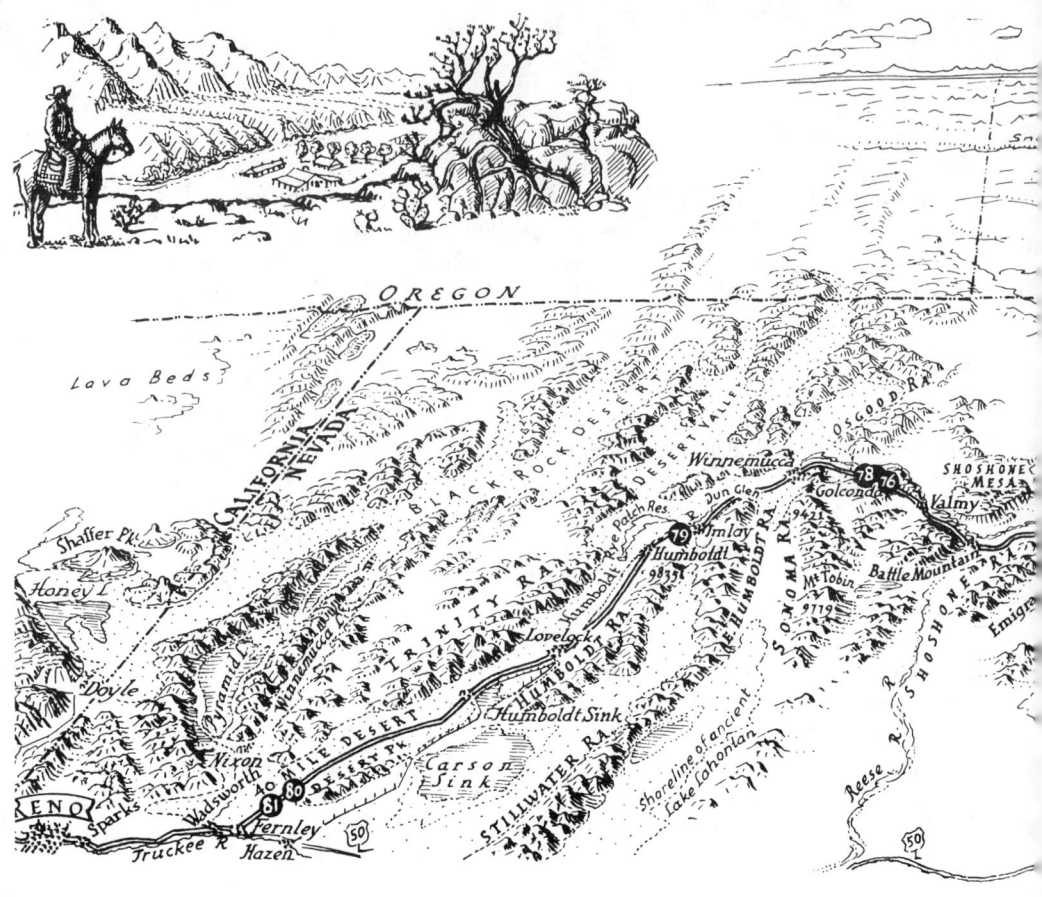

## CUTOFF AND CALIFORNIA TRAIL
## Salt Lake City to Reno

U.S. 40 between Salt Lake City and Reno not only reaches from one regional capital to another, but also crosses the width of the Great Basin. It passes, for about half the 530-mile distance, across the now dry beds of those ancient inland seas, Lakes Bonneville and Lahontan.

Scenically the region is scarcely second to any. Jagged ranges of moun-

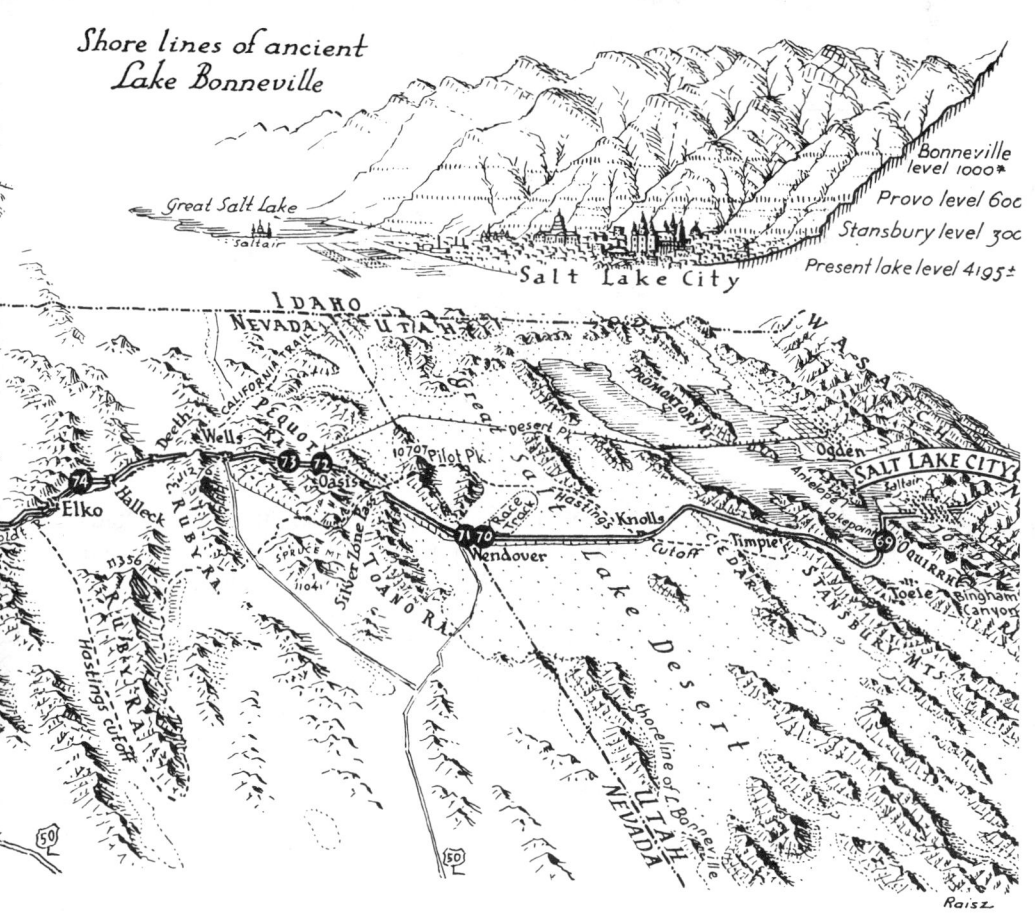

tains, sometimes snowcapped, trend north and south, and between them lie desert valleys, majestic in empty distances. Mirages distort the sky-line, and dust-devils blow on the salt-flats. Though the land itself is often a gray monotone, the light and color of the desert fill the atmosphere, and the clouds offer infinite variety.

As far as the little station of Oasis, Nevada, the highway roughly follows the so-called Hastings Cutoff.... In 1846 Lansford W. Hastings rode eastward, hunting for a shorter route by which the covered wagons could reach

California. Leaving the established trail at a point not far from the present town of Elko, Nevada, he sought to avoid the far northern swing *via* Fort Hall, in southern Idaho, and headed more directly eastward. He rejoined the old road near Fort Bridger, Wyoming.

Hastings was highly ambitious and his motive in thus establishing a new route was apparently to make himself an important man in the eyes of the emigrants, with the ultimate object, perhaps, of organizing them for an American revolution against the Mexican government of California. Although a poor trail-finder, entirely too prone to take chances, he was a good talker, and persuaded many emigrants to leave the established road.

The first party, guided by Hastings himself, got through without serious difficulty, by hard work and good luck. The second was the famous Donner Party. They too managed to get through, but they were forced to abandon some wagons in the desert, and they lost many oxen. The disaster that overtook them in the Sierra Nevada is to be attributed largely to the time they consumed in passing along the cutoff. This route, having one stretch of eighty miles without water, was certainly ox-killing, if not man-killing. A few of the Forty-niners followed it, but thereafter it may be considered out of use, at least westward of the little Mormon town of Grantsville.

The route of the overland stages and of the pony-express kept to the south; the transcontinental railroad, to the north. Only in 1907 was another route of travel established along the general line of the cutoff, when the Western Pacific Railroad was built across the Bonneville salt-flat, east of Wendover.

A few years later, with the automobile horns blowing ever louder, the officials of the Lincoln Highway began to map out their route westward from Salt Lake City. They selected one that approximated the line of the stage-road. The people of Utah, however, saw greater possibilities in a route that led more directly westward, close to the new railroad. A bitter controversy arose. The Lincoln Highway route was direct; though somewhat mountainous, it offered few engineering difficulties, and would serve for travel both to San Francisco and to Los Angeles. The Association put up money for the improvement of its road, and then charged bad faith when Utah determined that the so-called Wendover Cutoff should be the state highway.

Against this route, it could be argued that the mud flat was five hundred feet deep, without solid foundation, so that any road laid across it would sink and disintegrate in time of rain. This was exactly what happened to the first road, constructed in 1916.

Two psychological factors were probably responsible for keeping the people of Utah determined on the Wendover route. First, the flat offered a magnificent high-speed roadway, in dry weather. Second, a road across it paralleled the railroad, and during two generations the people of the West had come to look upon a railroad as something that a road naturally followed.

In the end the engineering difficulties were surmounted, and an enduring road was completed in 1924. The Association, then nearing the end of its activity, was forced to capitulate, and to reroute the Lincoln Highway.

To say that U.S. 40 follows the Hastings Cutoff is true in any strict sense only from Salt Lake City to a point near the little station of Timpie. There the route of the cutoff turns sharply to the south for a dozen miles, then bends back to cross the highway a few miles east of Knolls. From that point it continues northwestward, over the end of the Desert Range, to rejoin the highway once more at Silver Zone Pass. The modern route therefore represents an extreme straightening, and the old road is in places fifteen miles distant from the new one.

Near Oasis the cutoff, following its highly erratic course, swings off far to the southward, not to return to the highway until after the latter has joined the main California Trail. The stretch of U.S. 40 from Oasis to Wells, a distance of 27 miles, is thus unusual in having no history before the automobile period. This road across the Pequop Mountains was constructed to connect the already established road-system of Nevada with the newly built Wendover Cutoff.

At Wells, Nevada, U.S. 40 enters the valley of the Humboldt River, and joins the route of the California Trail, which latter it follows all the way to Reno, and beyond.

The Humboldt valley, broad and not excessively circuitous, offers the easiest route across the greater part of what is now Nevada. Since 1825 various parties of trappers had visited the region, and in 1833 one of these under the leadership of the renowned Joe Walker probably followed the course of the river, on horseback, clear to the sink. The Bidwell Party of emigrants, having abandoned their wagons, did the same in 1841. The

Chiles Party of 1843, led by the same Joe Walker, is believed to be the first to have taken wagons along the Humboldt. Very scanty records are preserved of this journey, and some of these seem to contradict others. Still, the weight of evidence points to the opening of a wagon-trail from "near the head of the north fork" (whatever stream may be so indicated) all the way to the sink. From that point however, the Chiles Party went south, and eventually had to abandon their wagons.

In 1844 another wagon-party boldly set out for California. Fortunately, we possess fairly detailed records of this important episode of American road-breaking, particularly in the reminiscences of Moses Schallenberger, who made the trip as a lad of seventeen, and when nearly sixty recounted its incidents. This is known, most commonly, as the Stevens Party.

This party is highly unusual in American annals in that most of its members were trying to escape from the United States. This majority consisted of recent immigrants from Ireland. They had not liked the Protestantism and anti-Irish feeling that they had encountered, and had therefore determined to migrate to California, which was Mexican territory and Catholic-dominated.

The party followed the well-established Oregon Trail as far as Fort Hall. There they separated from the other emigrants, and set off southwestward on their own—according to Schallenberger, eleven wagons, twenty-six men, eight women, and about a dozen children. It was a small company to face desert and mountains, and attempt a major work of trail-breaking.

At the beginning of the journey the Stevens Party hired as guide the well-known Caleb Greenwood, an old mountain-man. Historians have therefore generally assumed that the route of the California Trail, and of Highway 40, is to be credited to Greenwood. Certainly he should have had plenty of experience, for at this time (according to his claims) he was eighty years old, and had been in the West ever since 1812, when he had made the transcontinental journey with the Astorian party. Schallenberger's reminiscences, however, which have generally escaped the notice of historians, state that Greenwood's contract expired at Fort Hall, and that beyond that point he did not pretend to know anything about the country. Actually, like most emigrant companies, this one does not seem to have been strongly under the domination of any one leader. Probably their official captain, Elisha Stevens, deserves the chief credit.

Stevens is the sort of person about whom we know just enough to make us want to know more. Physically, he was of slim build, with a long neck, a curiously narrow and peaked head, and a great hawk-like nose. At the time of the journey he was forty-one years old. Born in South Carolina and raised in Georgia, he had already had some experience as a trapper, probably in the West. He was uneducated, but of "sound sense and great experience." In personal habits he was something of an eccentric, and generally lived as a hermit. Just how he happened to join an emigrant party is not known, and his election as captain—since he must have been an outsider—is probably to be credited to his actual abilities. Certainly, two members of the party later credited its success to his leadership. One of these attributed to him a whole sheaf of virtues: "cautious, polite, hopeful, courageous, prudent, plain, domestic, generous, attached to friends, firm, persevering and successful," and then added that he had mechanical ability and that on one occasion, when the way in the mountains seemed to be blocked, he prayed, and "it was given to him in a vision to keep right on up a ravine."

Stevens, Greenwood, and the others must have heard something at Fort Hall about the Walker Party of the year before, but Schallenberger never refers to them or to any trail left by them.

In any case, whether the Stevens Party followed wagon tracks for the whole or for a part of the distance, or whether they broke trail, they had no real trouble in reaching the Humboldt and in following along it. The stream furnished water; its meadows, plentiful grass. The party carefully kept on friendly terms with the Indians. Along the river the country was open and level. Schallenberger, years later, recorded of this part nothing more than: "The journey down the Humboldt was very monotonous. Each day's events were substantially a repetition of those of the day before."

The first difficulties arose when the party had reached Humboldt Sink, south of the present town of Lovelock. Here the river ended, and both to west and to south a desert stretched away. Even if there were tracks left by the wagons of the preceding year, these would have been of no comfort, for either the fate of those wagons was unknown or else they were known to have been abandoned. In effect, there at the sink, no one in the Stevens Party knew which way to go or whether the desert either to the west or to the south was passable.

In their difficulty they found an old Indian, chief of the local Paiute band, whom they came to know by the name of Truckee. Greenwood spoke a little Shoshone, and these Paiutes spoke a related language. The situation was about the same as when a tourist with a smattering of French attempts to talk with an Italian peasant. Signs and gestures, however, could help, and the general idea must have been fairly obvious. Doubtless, too, Truckee was as anxious to get rid of the white men as they were to leave. He drew them a map in the sand, indicating that they should go west. After fifty or sixty miles, the emigrants gathered, they would come to a river that flowed from west to east—a fine stream with trees and grass.

Three of the men prudently set out to explore on horseback, before they committed the wagons to the desert, taking Truckee along as a guide. Both his knowledge and his honesty were vindicated when the river was discovered, and in gratitude the emigrants called it by his name, which it still bears.

The party left the sink about October 8, having cooked two days' rations and filled all available vessels with water. They traveled with scarcely a halt from morning until twelve o'clock the next night, and then reached some small hot springs. After a halt of two hours, they resumed the march, and came to the river at two o'clock the next afternoon.

The people, having had sufficient water, had not endured any great hardship, but the oxen had suffered severely. They were given a two days' rest for recuperation on the fine grass beside the river.

Now, instead of following the Humboldt down, the party had to follow the Truckee up. The grade was against the ox-teams. Much worse, the Truckee came down through a canyon, and the stream swung back and forth from cliff to cliff. The wagon-train frequently had to ford from one side to the other.

When the party, after three or four days of this struggle, at last burst out of the canyon and saw before them a broad meadowland several miles wide, they must have been delighted, even though before them they saw the mountains looming up higher than ever. Farther to the west, in those mountains, the Stevens Party was to suffer still greater hardship and accomplish still more spectacular trail-breaking. For the moment, however, we may permit them a momentary respite in Truckee Meadows, where the city of Reno was later to be built....

The road thus pushed through in 1844 was followed by more emigrants the next year, and by a large number in 1846. It became the great road of the Forty-niners. Needless to say, it suffered innumerable minor relocations, and in 1849 was really double-tracked over most of its course, with branches on both sides of the river, and cross-overs and cutoffs, to accommodate, in some fashion, the incredibly heavy wagon-traffic of that incredible year.

During the sixties, after the establishment of the pony-express and of the stage-route farther to the south, along the line of U. S. 50, the Humboldt route became of secondary importance. The transcontinental railroad, however, naturally used the broad, continuous valley of the river, avoiding the many-summited route followed by the stage-coaches.

In the Dark Ages the old trail remained, in places shifted to parallel the railroad, still traveled by an occasional sheepherder's wagon. With the coming of the automobile the importance of the Humboldt route was again threatened when the Lincoln Highway Association selected the old stage-coach route. With the insistence of Utah on the Wendover Cutoff things were evened again, and the balance quickly settled in favor of U. S. 40.

U. S. 50, across Nevada, is a good highway over which the ordinary tourist can make just about as good time as he can by U. S. 40, and along which he can see rather more beautiful scenery. It is, however, a comparatively lonely road, with few chances for meals or lodgings.

U. S. 40 and U. S. 50 share the same pavement from a point twenty miles west of Salt Lake City clear to Wendover. From their point of separation, however, U. S. 40 is definitely "the main line." It follows the railroad, and so touches more towns. The heavy trucks keep to it, avoiding the many steep grades on U. S. 50. So too, the ordinary motorist naturally follows it, unless he has some particular reason for going the other way. There is about it, from Salt Lake City to Reno, nothing of that lonely quality that it displays in western Kansas, across most of Colorado, and in eastern Utah. No longer is it overshadowed by U. S. 30 to the north, or even in parts by U. S. 24, and 34.

. In general it is a broad two-lane highway, sometimes using the line of the old emigrant-trail, sometimes following along the railroad, sometimes leaving them both for a wholly modern relocation. Already its two lanes are beginning to seem crowded, even in the desert stretches. Through all the hours of the night the buses and the long-distance high-speed trucks roar through. Across Nevada U. S. 40 scarcely has a rival.

## 🞶 Great Salt Lake and Smelter

🞶 The view is southeastward from Black Rock, a dark mass once rising as an island from the water of the lake but within the last few years connected to the shore as the consequence of receding waters.

In the foreground is a little of Great Salt Lake itself, the incredibly gentle slope of its bottom being shown by the distance from shore to which a bather can wade without even wetting his knees. It is a Sunday morning, but the beach does not as yet display much activity.

At this point the Oquirrh Mountains (pronounced Oak'r) come close to the lake-shore, and all communications are squeezed into the bottle-neck between the steep slopes and the water. Beyond the beach can be seen—first, the little road to the beach itself; then, a line of railroad; next, a large pipeline; beyond that, U. S. 40, showing clearly in white; then, another railroad; finally, on the slope of the hill, U. S. 50, which joins with 40 just to the right of the picture at Lakepoint Junction. In addition, there are seven pole-lines.

More prominent even than roads and railroads is the great Garfield Smelter, hard at work producing copper from concentrates derived from ores mined in Bingham Canyon, farther south in the Oquirrhs. Three tall chimneys, painted black at the top for airplane warning, are pouring out smoke. Heavily charged with sulphur dioxide, this smoke has completely blighted the growth on the near-by mountains, although at best this is a desert range with scanty vegetation. The tremendous slag-dump shows as the long flat-topped dark mass at the left, connected with the smelter by a bridge under which the railroad passes.

Beyond the slag-dump, twenty-five miles or more distant, dim in the haze, rises the bold line of the Wasatch Mountains, the westernmost escarpment of the Rockies. Salt Lake City lies at the foot of this range, toward the left.

Ancient Lake Bonneville, of which Great Salt Lake is only a ten per cent remnant, once stood a thousand feet deep over the point at which this picture was taken. At a very recent period geologically, perhaps not more than thirty thousand years ago, the lake washed against the slope of the Wasatch. One of its shorelines shows clearly as a wave-cut terrace all along the slope of the Oquirrhs, beyond the three chimneys. At the point of the mountain to the left of the chimneys the terrace shows clearly, and two rocks jut up from it. These must once have stood a little offshore, as rocky islands with the waves breaking about them.

The Hastings Cutoff necessarily came through this bottleneck. Hastings himself, with the party he was guiding, was camped near Black Rock one night early in August, 1846, when James Reed, scouting ahead for the Donner Party, came into the camp. The Donner Party itself camped at the same spot toward the end of the month, and one of them, Luke Halloran, having died of tuberculosis during the day's journey, is buried somewhere in this vicinity.

## ⁷⁰ Salt-flat

🛡️40 For the benefit of anyone not familiar with salt flats, it may be stated that this picture was taken toward the middle of a blazing summer day and that there is absolutely no snow in it.

The view is eastward from a high rocky point, just east of the town of Wendover. The following picture was taken from the same point, by reversing the direction.

Even without being told that a town was near, the experienced traveler should be able to deduce it from two evidences. First the railroad yards indicate the vicinity of a town. Second, the billboards begin about a mile down the highway, and they are all on the left-hand side of the road, thus obviously placed to catch the eye of west-bound motorists, and advertise the business firms of Wendover.

The landscape is here again chiefly determined by Lake Bonneville. In fact, even the high point from which the picture was taken was once 800 feet beneath the surface of the water.

In the distance, about sixty miles away, the line of the Cedar Mountains is faintly discernible through the desert haze. The lake once washed high against their slopes.

The white line of the salt-flat seems to begin at the very foot of the mountains, but actually it is many miles this side of them. This flat, absolutely devoid of vegetation, a hard and level surface of salt, is about six miles wide and forty miles long, and represents the last part of this area of the lake to dry up. In a sense it is not even yet entirely dry, for in every winter it is likely to be covered with a foot or so of water.

Closer at hand the ground is a few feet higher, and so there is drainage toward the flat. For this reason some of the salt left by the evaporation of the lake water has been carried away, and a scattered growth of desert shrubs has become possible. Any minor depression, however, collects water, and therefore salt, and becomes so thickly encrusted that even the hardiest greasewood cannot grow there. Several such minor salt-pans, white as snow, appear just beyond the railroad. Precipitation is here down to well below five inches annually.

An amusing detail, showing the constantly rising standards of highway construction, is to be seen at the first bend of the highway. In the early twenties cars were still moving at comparatively slow speeds, and so the highway was run out almost to an angle. This actually meant the construction of a somewhat longer road, and therefore the use of more material. On the other hand, it meant simpler paper-work and surveying. Inevitably the curve had to be eased to accommodate the high speeds of modern cars.

The farther curve results from the necessity of avoiding the railroad right-of-way, which is broader in the vicinity of the yards. From that point on, the highway keeps as close as possible to the track. This was advantageous for construction, since the railroad had been built first, and materials were brought out by rail.

The highway parallels the track across the flat—with some minor deviations following a straight east-west line for almost forty miles. This is the second-longest interval between gasoline stations on U. S. 40, the longest being just to the west of Lovelock, Nevada.

# 71 Wendover

US 40 Wendover, on the Utah-Nevada line, began as a railroad town on the Western Pacific, but the railroad has ceased to mean much in its life, and it has become a typical desert highway-station. To tourists, truckers, and bus-passengers it is the natural stopping-place—for gas, or lunch, or a cup of coffee—in the 193-mile stretch between Salt Lake City, and Wells.

The importance of U. S. 40 in the life of Wendover may be seen by the way in which all businesses string out along the pavement. At the foot of the hill is a small motel; then, two service-stations, next, a garage; then cabins, another motel, and a third service-station. The large white building on the rise, to the left, is the chief hotel-restaurant, the "bus-stop."

As a highway-town, Wendover is all of a piece, except for the tightly built rows of war-houses, which hark back to the time when this was the supporting town for a bombing-range.

As usual in modern western towns, the scattering of houses is extreme. A car is now as great a necessity as a horse used to be. Only people who were assuming the use of automobiles would so have scattered a town.

As a highway-town Wendover has a special importance because it is at the separation, or the joining, of U. S. 40 and U. S. 50. In the foreground the four-lane stretch represents the two highways combined. Their point of separation is behind the low hills in the middle distance, but to the left the white line of 50 shows clearly as its heads southward toward Ely, and 40 goes off to the northwest between the two masses of black rock. The rivalry between the roads was at one time so intense that the towns along U. S. 50 hired an agent to frequent service-stations in Wendover, engage in conversation with westbound tourists, and encourage them to turn left.

Although this view is only the reverse of the preceding one, the general effect is startlingly different. Actually, all this country, as far as the Toano Range in the distance, was under the water of the lake. Even the mass of black rock at the right would have been deeply covered. At various times during the recession of the water, each of the higher points would first have stuck out as island and then gradually have been connected with the mainland as a peninsula. The notches on the dark rock are old wave-cut terraces, but on the whole few traces of the lake appear in the picture. Since the time of the desiccation the ordinary processes of erosion have obscured the old beaches. Moreover, on these higher slopes the water drained away and was not evaporated, so that the whiteness of salt shows only in one small spot at the left. The valley and the lower slopes are mostly sagebrush-covered. The higher mountains show the darker patches of juniper.

The cloud development indicates early afternoon. Rain is falling on the mountains to the right.

# 72 The Great Basin

The view west as U. S. 40 emerges from the windings of Silver Zone Pass may be considered typical of the Great Basin and therefore of Nevada. The name Great Basin, first given by Frémont, is accurate in that it indicates a large area lacking outlet to the sea; it is inaccurate in that there are actually many small basins, no single great one. As here, the ranges generally trend north and south, and empty sagebrush-covered valleys lie between.

The straightaway across the valley is about ten miles long. It drops away from the slope of the Toano Range, and rises again against the slope of the Pequop Mountains. The point of view is here at about 5500 feet; the valley floor is 500 feet lower; the highest points of the mountains, a little over 8000. The highway-station of Oasis is at the foot of the mountains near the end of the straight stretch.

The vegetation in the foreground is the typical sagebrush; on the mountains the darker tones indicate pinyon and juniper. Although the roadwork is twenty years old at least, one should note that the sagebrush has not been able to reestablish itself. This is an interesting indication of the slow rate at which vegetation spreads in dry country, and increases our confidence that the lines of old emigrant trails may actually still be followed even though they have remained untraveled for many years.

In this connection the present picture has its historical interest. The Hastings Cutoff crossed by way of Silver Zone Pass, and then struck across the valley toward the springs on the opposite side, following much the same route as the present road. At these springs the Hastings Party found some abandoned wagons of the Bidwell Party of 1841. The Donner Party also crossed through Silver Zone Pass, but at about the point from which the picture was taken, swung southwestward, taking a shortcut and reaching some springs farther to the south.

Along the center-line of the valley, north and south, runs the Nevada Northern railroad, serving the large copper mines of Ely. Some white spots among the sagebrush indicate the line of cuts along the railway.

This picture, seemingly taken by the simple process of setting up a tripod on the center-line of the highway and exposing the film, is the result of a number of favorable factors, and the author has not been able to equal it in twenty shots, taken at this and similar spots, at other times. The sun was low and squarely behind, so that the blackness of the highway stood out in contrast to the glaring whiteness of the cuts along the road. The telephone line, at this point, happens to be at an excellent distance from the highway, neither too near nor too far. The weak clouds of early morning (it was about eight o'clock) serve to outline the mountains without subordinating them.

**73** Pequop Summit

*Salt Lake City to Reno* · 251

(40) As U. S. 40, heading eastward, rises sharply in a sweeping curve toward Pequop Summit, it is well over a mile high. Even at this altitude, however, aridity predominates, and the scanty vegetation is of desert types. Sagebrush and bunch-grass dot the rocky slopes thinly. A few junipers, and an occasional pinyon-pine, have managed to strike root.

The juniper in the foreground is typical. Not more than fifteen feet high, it is nearly two feet thick at the base, and may be several hundred years old. At the right of the picture, the taller tree approaching more closely the form of a pine is a pinyon. Both these trees were of great importance to the desert-dwelling Indians, who depended upon pinyon-nuts for a valuable food supply, and used juniper for fire-wood.

The bold topography of the Pequops is typically far-western. The sheerness of the two cliffs is here caused by the jointed structure of the beds of massive rock, which causes them to break off vertically, whereas the beds of softer rock weather so as to form a slope. Such conditions could of course obtain in the East, also, but there the heavy rainfall tends to produce different effects, and in particular the tree-growth obscures and softens the profiles. The view here can be contrasted with that of Wills Creek narrows on page 96. In that view of the Appalachians many more cliffs would certainly appear, and the landscape would look much more rugged if the forest should be removed.

This picture was taken, as the small shadow beside the car on the highway would indicate, late in the morning. Nevertheless, the clouds have not developed greatly. At the moment they were actually beginning to form very rapidly, and to shift shape from moment to moment, as they drifted rapidly toward the south, from left to right across the background.

In spite of the comparative barrenness, either cattle or sheep can find a nutritious pasturage on the bunch grass of such mountains as these. The owner of the land to the south of the highway has not bothered to erect a fence, but a short line of fence appears to the north, filling the gap not protected by the cliffs themselves.

This is a part of the short section of U. S. 40 that has no distant historical background, but was constructed in very recent times, wholly as an automobile road.

# ⑦④ Sun on Highway

🛣️ Halfway between Halleck and Elko, U. S. 40 straightens out on a five-mile east-west tangent across rolling sagebrush-covered country. The view is a little to the north of east. The time is nine o'clock of a summer morning.

The sun is still low enough to make the bituminous-surfaced roadway stand out brilliantly in the view of anyone looking toward the east. Its heat, however, is not yet powerful enough to make clouds rise even over the mountains, and the skyline stands out clean-cut.

The mountains, their tops about twenty-five miles distant, form the East Humboldt Range, one of the higher desert ranges. Hole-in-the-Mountain Peak, their highest point, reaches an elevation of 11,276 feet. The highway, following the river, swings north around the end of the mountain, visible at the left.

At this point the highway has left both railroad and river to the south, and is cutting across a river bend. It has at the same time left the line of the emigrant road, and is here on a modern location. This is the only place in Nevada, except for a few streets, where U. S. 40 follows the lines of the land-survey, and thus runs directly east and west.

The topography is here more rolling than one finds along the river. The predominant growth is sagebrush, as is to be expected in "The Sagebrush State." Actually, this plant is not nearly as common, even in Nevada, as the ordinary passer-by is likely to think. Other desert shrubs give much the same general effect, and the tourist is likely to think that he sees sagebrush when he is actually looking at greasewood, or shadscale, or something else. Sagebrush is even, in its way, a somewhat fastidious plant. It has at least four requirements—a dry climate, a cold winter, a fairly deep soil, and a low salt content in the soil.

In this picture, for instance, a spot just to the left of the road is bare of sagebrush, and another spot in the middle distance to the right of the road. In this second location the difficulty is obviously too much salt, for the whiteness of the surface indicates alkali.

The road itself is here a good example of the broad two-lane pavement that is characteristic of U. S. 40 across Nevada. Having just come over the crest of a hill the highway shows the "No-Passing" line at the center. It has also been especially widened with an additional shoulder, which shades back to normal width at the bottom of the hill.

Two pole-lines, like the highway, take the shortcut—unlike the railroad—not caring about the steeper grades thus necessitated. In fact, the unnecessary way in which the pole-line crosses the highway at the bottom of the hill, probably indicates that it was here before the highway.

## 75 Emigrant Pass

🛡40🛡 The names Emigrant Pass and Emigrant Springs are scattered widely over the West. Why these people were always conceived as emigrants and not as immigrants is a problem to be referred to the folklorists, or perhaps to the psychiatrists. Is it to be taken as a symbol that Americans were always conceived as escaping from something rather than as attracted toward something?

At this particular Emigrant Pass the highway-engineers have agreed with the scouts for the covered-wagons by leaving the Humboldt River just west of Carlin. At the expense of climbing a spur of mountains, wagons and automobiles alike save some miles of distance by cutting across a bend of the river.

The view here is from west to east, looking back at the pass, which is nothing more than the mild sag in the skyline of the hills.

Although the old road can be easily followed on the other side of the hills, it has on this western slope been wholly obscured by the construction of two later highways, demanding a considerable amount of cut-and-fill.

There are two springs, distinguished by the emigrants as Upper and Lower. The latter is clearly marked in the picture by the dark mass of trees. The trees around the more distant spring are hidden behind a roll of the ground. Since I have found no early mention of tree-growth at these springs, I would assume that they originally showed up only by the dark ring of bulrushes around them, and that the cottonwoods were brought in later by white settlers.

The second section of the Donner Party camped here, early in October, 1846. With their usual bad luck or bad management they almost lost three wagons in a grass fire, and during the night had a shirt and two oxen stolen by the local Paiutes. Not infrequently, even in these days, a traveler loses his shirt while crossing Nevada.

Since this is the only water-supply along a considerable distance of the highway, service-stations and overnight cabins have been established at the springs.

The rolling, easy contours of the terrain are indicative of what is known, geologically, as the old surface of Nevada, that is, a terrain that was worn down to comparatively gentle slopes some ten million years ago, and since then has suffered little change, thus offering a contrast to the jagged peaks of the more recently uplifted ranges, like the Toanos and Pequops. Such country usually has fairly deep soil, and therefore, as here, supports a good growth of sagebrush interspersed with grass, and offers good grazing.

In color-tone as in relief, the old surface is prevailingly lacking in contrast. The gray tones of the picture present an accurate rendering of the country.

In such country summer is the most colorful season because the cumulus clouds provide its most scenic feature. In this picture, taken at twelve-thirty on a June day, the clouds have just risen sufficiently to set off the line of the hills, and thunderheads are beginning to poke up.

## 76 Tie-House

⑩ A few miles west of Valmy a spur of mountain juts out close to the highway, and on its slope a miner (his claim is a little way up the mountainside) has constructed a typical Nevada house. It is built in the log-cabin tradition, but in this treeless country no logs are available and old railroad-ties serve instead. It has a tarpaper roof, almost flat, since there is little rain or snow to be feared. Fire is a much greater danger, and so the stovepipe has been carried up high above the roofline, and capped to prevent the escape of sparks. To avoid the winter winds the house has been faced eastward, and exposes only its end to the north.

In this country water is the great determinant, and the location of this house is dependent on the little spring that breaks forth from the slope of the mountain just at this point. The settler has run a barbed-wire fence around the spring, partly as a mark of possession, and partly to be rid of the nuisance of wandering cattle coming there to drink, and polluting the water. The growth inside the fence, where it is protected from grazing, is markedly different from the growth on this side of the fence. Only a few clumps of tall rye-grass have survived outside, but inside it is the predominating growth.

The Lombardy poplars, planted to the south and west of the house for shade, thrive where there is a little seepage of water from a spring, and have become the most characteristic introduced tree of this region. The two small pine-trees, to the left, are much less characteristic, and do not seem to be thriving.

The miner has, as usual, shown no esthetic feeling. His dump for tin cans is directly in front of the house, and he has chosen an equally prominent location to deposit the wreck of a car.

To the left, the two sheds are even more primitive than the house, being roofed merely with earth. Such a roof offers good protection against snow and cold weather, and in a country where rain may be limited to two or three hard showers in the course of a season, it is also a good enough protection against rain. The sheds, probably constructed for the protection of horses, suggest—as does the size of the poplar trees—that a location was made here before the automobile period.

Between house and highway the ground is highly alkaline and there is little growth. This is probably the bed of one of the many small lakes that flourished in the more rainy past. A small pool of water, marking another spring, appears at the left.

Beyond the highway, particularly toward the left, the growth gradually becomes richer as the Humboldt River is approached.

The line of the river itself is not at all conspicuous. Two railroads here parallel it, and the white lines of their embankments are faintly visible—the Southern Pacific on this side of the river, the Western Pacific on the other.

In the distance the mountains of northern Nevada, heavily shadowed by a storm cloud, are silhouetted sharply against the brighter sky to the north of the local afternoon thunderstorm. Rain is falling upon the mountain to the right.

**77** Playa

(US 40) Only a few miles west of the tie-house the highway crosses the flat expanse of what is locally known as a sink, but is called by geologists a playa. This latter term, meaning *beach* in Spanish, was doubtless first applied to the ancient beaches of extinct lakes, and was later extended to cover the whole lake-bed. Bonneville and Lahontan were the giants of these ancient lakes, but scores of smaller ones also existed.

The one here pictured is too small to have a name. Its history presumably falls into three divisions. In the most rainy periods—associated with the advances of the ice-sheets from the north—it was fresh, and overflowed into Lake Lahontan, which stood at a slightly lower level, even at its time of greatest expansion. When rainfall became less, the little lake ceased to overflow; shrinking back, it became brackish and then salty. Finally, in the third and still existing period, it dried up, leaving a deposit of salt.

If a heavy thunderstorm breaks over this area, the lake bed will collect the run-off from the mountains around it, and will again show for a few days or weeks, a broad sheet of water—until seepage and evaporation again do their work. The highway, it is to be noted, has here been built up well above the level of the playa to prevent such occasional flooding.

From the point at which the picture was taken the ancient shoreline can easily be seen—where the solid darkness of sagebrush, at a slightly higher level, replaces the largely bare ground.

A playa can generally be recognized, as here, by its very light colored soil, often salt-crusted. On account of the high salt-content, sagebrush does not grow in a playa, and the common plant is greasewood. The only other growth here showing is a very little salt-grass, visible as a slight fuzz on the ground at the left. The growth of greasewood is characteristically scattered, as here, probably because each plant, seeking all possible moisture, sends its long roots out into the intervening spaces, and thus prevents the establishment of seedlings except under especially favorable conditions. Another interesting characteristic of greasewood, shown at the right, is its tendency to grow on the top of low mounds, the origin of which is difficult to explain.

The telephone-line here does not follow the road, probably because this is in a newly established location. Only the poles of a power-line march across the sagebrush.

The tie-house shown in the preceding picture is located just around the point of the mountain at the right.

**78** Golconda Summit

Salt Lake City to Reno • 261

**(US 40)** At Golconda Summit the highway crosses a low mountain spur, thus cutting across a bend of the river and leaving both the railroad and the emigrant trail, which follow the easier grade along the river. At the station of Golconda, shown by the usual dark line of trees in the distance, the highway rejoins the railroad.

U. S. 40 here swings to the left in a magnificent curve, descending from the summit. Then, reversing to the right, it may be seen crossing the valley, and can even be followed with the eye clear beyond the town. In all, about six miles of road are in view.

Although here the highway does not go back to the emigrant trail, three older roads can be seen in the picture. A recently abandoned highway shows at the extreme right. At the point where the farther trailer appears, this road joins the present highway; from that point on, they coincide. The faint line of a somewhat older but still modern road can be seen leading to the concrete culvert at the left. Slightly to the right of the culvert, the trace of what appears to be a much older road is cut out of the side of the slope. This continual relocation of roads is typical of the automobile period, and quite possibly the excellent highway here seen may itself become obsolete in another generation. If so, it will leave another scar on the landscape for some future archeologist to decipher.

The terrain is typically Nevadan with a broad sagebrush-covered valley ending against rocky and jagged mountains. These mountains form the Sonoma Range, with the highest point in the picture about 7650 feet.

Lake Lahontan extended just about as far as Golconda, and if this picture had been taken 20,000 years ago, we might see a little water in the distance at the right. From Golconda U. S. 40 runs in the ancient lake bed clear to Wadsworth, Nevada, and a little distance beyond, a total of about 150 miles.

The clouds represent an early afternoon in summer. Thunderheads are rapidly building up, and at one point on the distant mountains a little rain is apparently falling. These clouds may be compared with those in the next picture, taken several hours earlier on the same day.

A note of alarm may be sounded over the appearance, even in this desert country, of a billboard.

## 🔴79 Imlay

🛡40 U. S. 40, paralleling the railroad at about a quarter-mile distance, here runs into the northeast. To the left, among the massed cottonwoods, is the little town of Imlay. Having grown up as a railroad town, Imlay still turns its back on the highway. Its cottonwoods are a mark of Nevada towns, which can frequently be seen looming up as dark masses in the distance.

The highway is here typical of much of U. S. 40 across Nevada, of two lanes, but almost wide enough for four lanes, and in addition having wide shoulders. It seems here to be heading toward a pass in the range ahead, but actually it swings far off to the north, by way of Winnemucca, around the point of the range on the extreme left.

These Humboldt Mountains, almost bare of trees, form a typical north-south desert range with their highest peaks being Dun Glen (7430 ft.) to the north and Auld Lang Syne, about the same height, slightly to the left of the highway. Both names are probably reminiscent of mines located by sentimental Scottish prospectors. South of Auld Lang Syne, just at the right of the picture, the ridge falls away to Natchez Pass, named probably for an Indian whose name is usually spelled Naches. On account of the clarity of the atmosphere the mountains seem closer than they are, Dun Glen Peak being about twenty miles distant.

The cloud-formation indicates late morning, when clouds form over the mountains rather than over the valleys, and have not as yet developed vertically. Only incipient thunderheads have formed as yet, and any rainfall is several hours off.

As often in the open reaches the pole-line forms a conspicuous feature of the landscape. Here we have the old transcontinental lead in all its magnificence. The big poles support four double-braced cross-arms with a total of forty wires. Their arrangement into pairs, and the use of double pairs of insulators to provide cross-overs, indicates that most of these are long-distance lines. Although there are only forty wires, the manipulations of electrical engineering allow, not a conventional twenty circuits, but 49 voice-channels (including two for radio-programs), and 39 telegraph channels.

Regretfully one must add that this great pole-line is already secondary, and will be pulled out and junked as soon as the telephone-company can get far enough ahead in its work to find the time. The great bulk of the transcontinental calls are now carried on a buried cable. Significantly, this cable departs from the line of the highway and railroad in this area, and takes the short-cut from Battle Mountain across to Lovelock—a much shorter route which the highway will very likely follow at some future time, the greatest relocation that is possible in the present route of U. S. 40.

# ⑧⓪ The Forty-Mile Desert

🛣️40 In its name the Forty-Mile Desert records the distance of the dreaded dry drive from Humboldt Sink to Truckee River. In using this name the emigrants, quite rightly, ignored the small hot springs in the middle, for at these they found no grass, and found water only in small quantities and of doubtful quality.

The picture was taken looking southeast from a small plane at an altitude of about one thousand feet above the ground, about three miles east of the hot springs. In the distance, about thirty-five miles away, the Stillwater Range supplies the skyline. Closer, to the left, the point of the Humboldt Range extends south into the desert. At the right, the low rounded summits of the Desert Range stand out darkly. The cloud-bank covers Humboldt

Sink, to the left, and the much larger Carson Sink, in the center and to the right.

Lake Lahontan, at its highest level, covered most of this area. The highway, if it had then been in existence, would have been more than 300 feet under water. The Stillwater and Humboldt ranges were long mountainous peninsulas—probably covered with pine forests—extending out into the lake from the north. The points of the Desert Range stuck up as small islands. The height of the lake-surface was, indeed, not so very far from that of the low-lying cloud-bank shown in the picture. By imagining the line against the base of the Stillwaters to represent water instead of cloud, the observer can have an idea of the lake as it once was. This expanse was the widest stretch of open water on Lake Lahontan, which covered a large part of northwestern Nevada, but was everywhere divided into long fiord-like arms.

The horizontal lines on the lower slopes of the Desert Range are the beaches left by the lake as it receded. The white patches of soil are also an evidence of the lake, because the fine light-colored silt left on its bottom is easily eroded either by rain or wind.

This portion of U. S. 40, here seen absolutely devoid of cars, is part of the longest stretch anywhere on the road to be completely without service-stations. From Lovelock to the hot springs, a distance of 41 miles, there are no supplies.

The old road intersecting the highway at an acute angle in the right-hand lower corner of the picture is the California Trail. In all probability it represents the actual line over which the wagons of the Stevens Party were driven in October, 1844. Although it shows by its windings the characteristic adaptation of an emigrant trail to the terrain, it is straighter than such roads usually are. On our flight (November 18, 1950) we picked up the track of this old road at a point several miles to the west of the Hot Springs, and followed it continuously to this point, and still farther, until we were frustrated by its disappearance under the cloud-bank.

This cloud-bank is as little typical of this area as anything that can well be imagined. It indicates that the ancient conditions have to some extent been re-established, and that large parts of Humboldt and Carson sinks have temporarily been covered with water—a result of the almost unprecedented rains that fell during the early part of November, 1950.

## ⑧¹ Lahontan Story

🛣️⁴⁰ A few miles east of Fernley, Nevada, the observer stands on the point of a rocky hill about a hundred feet above the flats, and looks out eastward across the highway, which here trends from northeast to southwest.

In the foreground, on the slope of the hill, the rocky slope is dotted thinly with bunch-grass and small desert shrubs. Lower down, the concentration of alkali in the soil is heavier, as the whiteness shows. In this lower zone the chief growth consists of scattered greasewood. Through this zone, skirting the hill, runs the transcontinental telephone lead, seen more closely in the second picture preceding.

In the middle ground, the concentration of salt is so great that the soil is almost snow-white, and even greasewood can grow only in a few spots. The dark area on both sides of the highway at the right is a small sink where water stands during part of the year and where plants that are favored by salt-water are enabled to grow.

Beyond the white of the salt rises the typical "old surface" of Nevada —gently rolling, monotone in color. High above these hills towers the piled-up mass of fracto-cumulus clouds, characteristic of the desert country in winter and spring. These clouds are the last remnants of a Pacific storm which has swept in, drenching California and the western slope of the Sierra Nevada, but has expended itself before reaching Nevada.

To work back behind the present, one should notice that between the highway and the hill run three very faint wheel tracks. The two that are nearest to the poles are probably the marks of the trucks which were used in the construction of the pole-line, and are still used for its maintenance.

The line of tracks halfway between the poles and the highway, visible at the left, almost certainly represent the California Trail. Whoever guided the Stevens Party in 1844 would have kept as close as possible to the point of this hill in order to save distance, and this track can be followed clear to the hot springs and beyond.

Much of the Lake Lahontan story is here revealed by the heavy deposit of white salt and by the lines of the old beaches showing on the dark hill across the flat. At the time of the highest water-level the hill from which the picture was taken was a part of a very large island. The opposing hills formed a smaller island. The expanse where the salt-flat now extends was a strait, about four miles wide, connecting two parts of the lake. The waters of the lake began to fall, but the progress was not steady, and at times the surface remained constant long enough to wear a beach on the mountain-sides. After the lake had receded considerably, a ridge at one end of the strait was exposed, and the strait became a bay. Soon, however, a ridge at what had been the other end of the strait was also exposed, and the bay became a lake, some ten miles long and four miles wide. This smaller lake gradually shrank, and its water grew saltier. With the final desiccation, some thousands of years ago, the salt was left exposed on the surface of the flat, as we still see it.

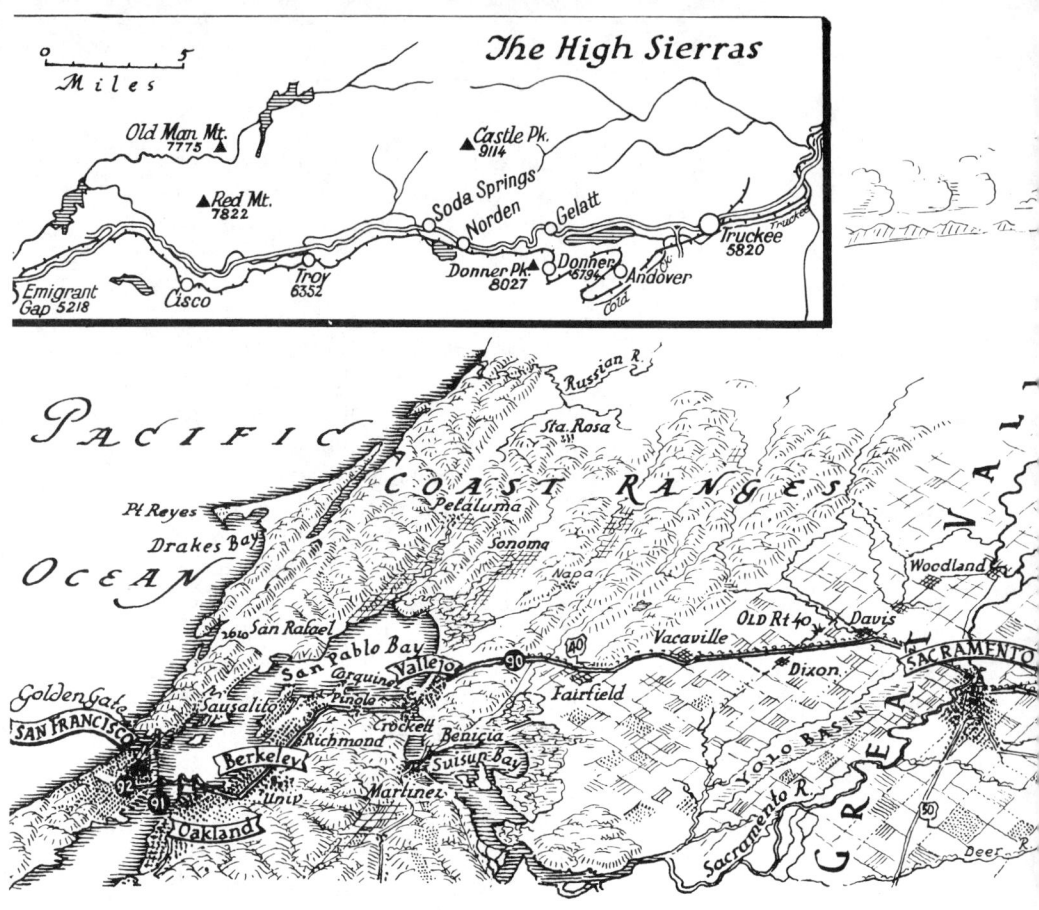

## TRUCKEE ROUTE

# Reno to San Francisco

In its westernmost sector U. S. 40, after a few miles in Nevada, crosses the breadth of California. Technically, it remains within the Great Basin until it passes the summit of the Sierra Nevada; the streams of this slope flow eastward and never reach the ocean. Actually, from Reno westward, the scenery changes, and the traveler feels that he has left the desert

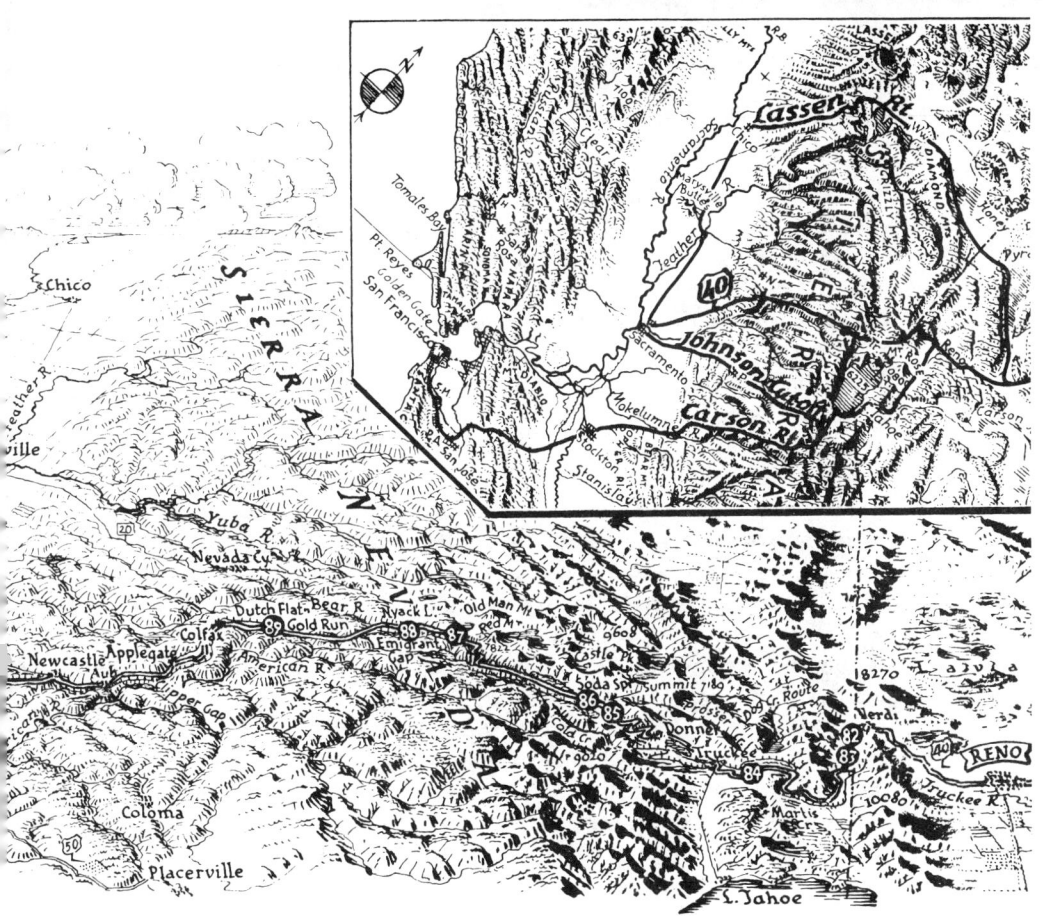

behind. There is little sagebrush; the mountains begin to be pine-covered and to rise higher. The total distance is 228 miles.

Topographically, this sector crosses the three narrow north-south bands that constitute the state of California for most of its length. First, toward the east, is the Sierra Nevada; next, the Great Valley; finally—and finally, also for the continent—the Coast Range.

For scenery, the Sierra Nevada ranks high. The Great Valley, on the contrary, is probably as uninteresting as any district on the whole road, being as flat as the High Plains, yet lacking their illimitable sweep, and seldom displaying much effect of light or cloud. The Coast Range varies between great beauty in winter and spring when the rounded hills are green and flecked with wild flowers, and dusty drabness in the late summer and fall when the hills are dull and brown. In between, during early summer, they show the rich golden color of ripened grass, spotted with the dark green of live-oaks.

The history of the road across the Sierra Nevada begins with the Stevens Party.... From Truckee Meadows they pushed on, fearing the approach of winter. Already it was mid-October, and snow had fallen. Hopefully pressing westward through wholly unknown country, they entered the upper canyon of the Truckee, thus keeping to the route of U. S. 40.

In spite of their worry about winter, they made only slow progress. The canyon grew narrower; the river swung back and forth so often that one day they crossed it ten times in traveling a mile. Heavier snows fell, covering all the grass, and only the rushes that were tall enough to stick up through the snow were left for the oxen to graze on; overworked and underfed, they grew steadily weaker.

A dilemma arose, when the weary emigrants came to a point where the stream forked, and no one knew which fork to follow. According to Schallenberger, a council was held and a decision reached after deliberation. According to another testimony, all was accident and misunderstanding.

In any case, one group of four men and two women followed up the main stream on horseback. After many adventures and hardships, they reached California safely, but their story has no connection with that of U. S. 40.

The main body of emigrants took the wagons about a mile up along the smaller stream, now called Donner Creek. There it was decided to abandon some of the wagons, and three young men, Schallenberger among them, volunteered to stay and guard the property from Indians through the winter.

The others then went ahead with six wagons. They worked their way for two miles along the north shore of a lake. In a little meadow, a quarter-mile beyond the head of the lake, they must have made a halt, and what they saw ahead was enough to appall anyone. A thousand feet high, two-feet deep in snow, so steep that the little streams often came leaping in cascades—a rugged granite mass blocked the way. Here the pass was only a somewhat lower spot in the range, and to reach that spot was really to ascend a mountain-side. Now certainly was the time for Captain Stevens to use his mechanical ability or to see a vision.

Doubtless some of the men went ahead to reconnoiter, but the known details are presented in a few sentences:

> All the wagons were unloaded and the contents carried up the hill. Then the teams were doubled and the empty wagons were hauled up. When about half way up the mountain they came to a vertical rock about ten feet high. It seemed now that everything had to be abandoned except what the men could carry on their backs. After a tedious search they found a rift in the rock just about wide enough to allow one ox to pass at a time. Removing the yokes from the cattle, they managed to get them one by one through this chasm to the top of the rock. There the yokes were replaced, chains were fastened to the tongues of the wagons, and carried to the top of the rock where the cattle were hitched to them. Then the men lifted at the wagons, while the cattle pulled at the chains, and by this ingenious device the vehicles were all, one by one, got across the barrier.

Thus the Stevens Party surmounted Donner Summit at just about the place where the highway still crosses it. To anyone looking at the pass today, the feat seems amazing to the point of being incredible.

The party took their wagons on, some twenty miles, still along the general line of U.S. 40. Then the able-bodied men, leaving the others encamped with the wagons, went ahead on foot, driving the oxen, expecting to return very quickly with supplies from the California settlements. As usually happened with emigrants, they underestimated the distance and the difficulties of travel; before they could return, the women and children were reduced to eating hides.

In the meantime, to the east of the pass, the three young men had built themselves a crude log-cabin, but they became panicky as the snow deepened. They made themselves make-shift snowshoes, and set out, early in December, to escape over the pass. Schallenberger became exhausted, and returned to the cabin, to take his chances there.

He had little food, and could shoot no game. Rummaging about, however, he found some traps. With these, in spite of the snow, he managed to catch coyotes and foxes. He lived on the foxes, finding them more palatable, and kept the frozen bodies of the coyotes as a reserve. Toward the end of February one of the party named Dennis Martin came back to aid him. Martin, a Canadian, understood how to make and use snowshoes. Thus aided, Schallenberger crossed the mountains.

In the end both the people and the wagons reached California in safety, although the wagons were not taken all the way through until well on in 1845. The actual completion of the California Trail must therefore be assigned to that year.

As a drama the story of the Stevens Party is second to that of the Donner Party. The earlier emigrants, however, were the more skillful and energetic, and also perhaps the more lucky. The greater fame of the Donner Party, however, has attached their name to Donner Lake, Donner Pass, and other features, and even U.S. 40 in that area is sometimes called the Donner Trail. Actually, however, the Donners came late, and in California are of no historical importance aside from themselves. The only road they established was across the Wasatch Mountains in Utah, where their name is not preserved. The situation is thus shot through with ironies.

The opening of the California Trail by the Stevens Party—the conquest of both desert and mountain—is a major triumph of American pioneering. As far as Humboldt Sink, these emigrants of 1844 were not the first to travel the route, and they were probably not the first to break a trail for wagons. From Humboldt Sink across the Sierra Nevada, however, they not only performed the remarkable engineering feat of taking wagons through, but they were also, so far as is known, the first explorers. From the Forty-Mile Desert in Nevada, to the Yuba River bridge near Cisco, California—a distance of 150 miles—U.S. 40 follows the general line of their exploration, and for much of the way covers the wheel tracks or is very close to them.

The later history of the Truckee Route, as it came to be called, is varied. Because of the great difficulties it was soon relocated in places. Two of these relocations are important, although neither of them in the end proved permanent. In 1845 Greenwood and others, returning east on horseback, worked out a circuitous route through Dog Valley, to avoid Upper Truckee Canyon—between Truckee and Verdi—where the Stevens Party had undergone such great hardships. Some time later, perhaps in 1846, an easier crossing of the summit was discovered by way of Cold Creek, the next canyon to the south.

The difficulties of the Truckee Route made emigrants try other approaches to California. The first of these was the Lassen Route, which swung far to the north. Emigrants who went that way, however, generally decided that it was an out-of-the-frying-pan-into-the-fire choice, and after a few years it was little used. On the other hand, the Carson Route, which was established in 1848, well to the south, took over a major part of the gold-rush migration in 1849, and relegated the Truckee Route to the secondary position. It came to be less and less used during the next few years, as still other routes across the mountains were developed. The chief of these was Johnson's Cutoff, opened in 1852, along the line of present U. S. 50. This became the famous stage-road between Placerville and Virginia City. As a result of these and other developments the Truckee Route must have been almost unused during the early sixties.

Donner Pass, however, proved the best suited for the construction of a railroad, and the officials of the Central Pacific saw it to their advantage to build a toll-road ahead of the railroad. This led to the Dutch Flat and Donner Lake Road, which was opened to traffic eastward from the railhead at Clipper Gap, near Applegate, in June, 1864. It eliminated the relocation by way of Cold Creek, and brought the crossing back to Donner Summit. Operating in conjunction with the railroad, this road carried heavy traffic, until in its turn it was killed by the completion of the railroad, five years later. It remained passable, however, and was the immediate basis from which U. S. 40 was later developed.

The long period of eclipse lasted until 1909, when the legislature appropriated funds for the old road's improvement. At that time it was reported "in such abominable condition that you could scarcely call it a road." In the next decade, however, only minor repairs were effected and I remember it in 1920 as still a narrow dirt road, with sharp turns and steep grades, not much differing from what it must have been in the sixties. Its sudden drop-off on the eastern face of the pass was breath-taking in beauty, but also heart-stopping, as one looked at the narrow and plunging road, corkscrewing off to some bottomless nowhere.

From 1924 on, there has been constant and rapid improvement, with many relocations, and the road is now an excellent two-lane highway, heavily traveled. In 1925 the Dog Valley detour was abandoned, and the highway opened through Upper Truckee Canyon. The elimination both of this relocation and of the Cold Creek crossing of the summit in favor of the original route is a striking testimony to the unswerving determination with which the Stevens Party pressed on directly toward the west.

During World War II the strategic importance of U.S. 40, as the best crossing of the Sierra Nevada, was suddenly recognized, particularly because of the ammunition depots located in the Nevada deserts. Partly upon strategic grounds the construction of a four-lane highway is being strongly urged.

From the western end of the Dutch Flat and Donner Lake Road, U.S. 40 merely follows the route of post-gold-rush roads and of the transcontinental railroad to Sacramento. From there to San Francisco also the route offers little of historical interest. The tule swamps and the arms of the bay made land-transportation difficult, and the Sacramento River furnished the natural means of early communication. The building of the railroad in the sixties rendered a road of even less importance, and so there is little to record before the automobile era.

There was indeed an early road, on a generally northeast-southwest bearing, between Sacramento and Benicia. This road, though marked on a map of 1849, seems to have disappeared completely, as the country was settled and private holdings in the level valley were laid out according to the section-lines. The only road known to the early automobilists ran around the corners of the sections, taking a zigzag course with right-angled turns, thus adding miles of distance. This section-line road was gradually improved, until it became an important highway. Only after World War II was it abandoned; in its place was established a four-lane freeway, following a direct course again, and at places probably approximating the original pre-section-line road.

The low Coast Range, broken by easy passes, raised little difficulty. The real obstacles between Sacramento and San Francisco were the arms of San Francisco Bay. To avoid the first of these, Carquinez Strait, the Lincoln Highway was actually taken by a roundabout route through Stockton. But a car-ferry solved the problem, and even before 1920 the direct route, the Victory Highway, had become the more important one. In 1927 completion of the Carquinez Bridge, almost a mile long, enabled motorists to make the run clear to Oakland. Ten years later the Bay Bridge, eight miles long, two-decked, hailed—and perhaps rightfully—as the greatest of all bridges, spanned San Francisco Bay itself, and allowed automobiles at last to roll into San Francisco over U. S. 40.

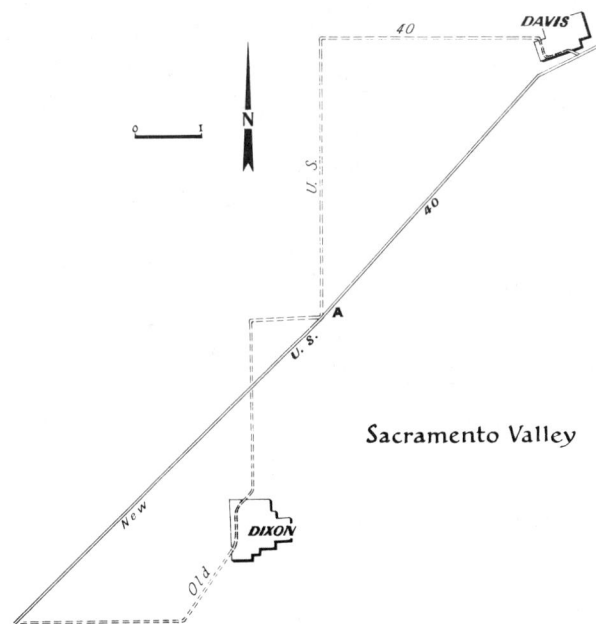

## ⑧² First Pines

🛣️⁴⁰ Half a mile west of Verdi, Nevada, U. S. 40 crosses the Truckee River. If the road continued straight ahead for a few hundred yards it would enter California. Swinging sharply to the left, however, it follows the course of the river, and stays within Nevada for three miles farther.

The direction of view is southwest. The bridge is at an altitude of about 4900 feet. The mountain rampart beyond, representing the first full-scale barrier raised by the Sierra Nevada, rises to a general altitude of about 8000.

Except for a few isolated trees in Verdi itself, these are the first pines to be encountered along U. S. 40 after its long desert sector. Even at this point there is scarcely anything that can be called a forest. The precipitation (twenty inches) is plenty for pines, and their scanty growth is to be attributed to the malign influence of civilization. Lumbering, followed by fires, has undoubtedly denuded these mountains, and in fact early emigrants noted pines at least ten miles farther east, in an area where they are not now growing. The almost complete openness of the ridge at the left indicates a fairly recent fire.

Individual trees show the effects of fire. The tall tree in the center of the picture is merely a trunk for about half its height, undoubtedly the result of the burning off of the lower branches. Such a tree has now become almost fireproof. The bark at the base is thick enough to protect the vital inner layers from the heat of any ordinary fire; the branches remaining are so high above the ground that they cannot be ignited; in such thin growth there is no possibility of a crown-fire jumping across from tree-top to tree-top.

Alders and cottonwoods, now leafless, grow along the river. The bench between the pine trees, and the dry slope beyond them, are covered with a scattered growth of sagebrush, dotting the snow.

The January scene shows the scanty snowfall of the eastern slope of the Sierra Nevada, which faces away from the ocean. Nevertheless heavy snows fall occasionally, and for the guidance of the snow-plows the road is lined with tall snowstakes, such as that one just to the left of the marker.

Near this spot, actually about at the village of Verdi itself, the emigrant trails split. The original route of the Stevens Party in 1844 followed the river, as the highway itself does. The route established in 1845 turned sharply to the right, broke through this opposing mountain barrier by a lower gap, and rejoined the line of the highway near the town of Truckee.

The dark cloud over the line of the mountains is a common local feature of winter. It indicates that a storm has swept in from the ocean, and that most of northern California is receiving heavy rain, with snow on the mountains. The storms, however, usually exhaust themselves on the western slope. At this very time a storm was raging on Donner Summit, and cars had to fight their way through, in convoy, against blizzard conditions.

## ⑧³ Truckee Canyon

🛡40 The Sierra Nevada near its eastern margin, close to the California-Nevada line, lacks the beauty and grandeur that it attains farther to the west along its central ridge. In Truckee Canyon the rocks are dingy volcanics, scaling off into long talus slopes and producing a somewhat tumble-down effect.

The forests are scarcely better than in the previous picture. Primevally this slope must have been covered with a fine growth of pine. The coming of the white man, however, brought both lumbering and forest-fires, and the combination proved too much for the recuperative power of the region. As a result the desert, at most only a few miles to the east, has begun to move in, and stretches of sagebrush have become more and more common. Considered individually, the trees are rich in color and healthy, but they are too often swept by fire to be able to mass together into a forest. The bare slope to the left of the road shows the actual mark of a fire that must have burned with a few years.

Actually, along a main highway and railroad fire-hazard is so greatly increased that, unless there is a vigorous attempt at reforestation, the prospect is lugubrious and the desert will probably continue to advance.

Here, where Truckee River cuts its way through the mountain barrier shown in the preceding picture, the canyon offers the easiest route, and communications bottleneck through it. In the foreground extends a three-wire, high-tension, transmission line. Next comes the highway itself. Across the river the railroad—doubletracked, but with the tracks at different levels—also makes use of the pass, and is flanked on both sides by pole lines.

Historically, the picture is of interest as showing the canyon up which the Stevens Party struggled with such difficulty in October, 1844. No trace of their wagon-tracks still remains, but one can deduce their route without much difficulty. Where the last stretch of water can be seen, they were probably forced to cross the stream, from right to left, because of the slope around which the railroad-tracks now run. After passing along the little flat, where the highway now runs, they would again have crossed, from left to right, and would then have reached the spit of level land below the railroad tracks. This spit would have brought them close to the river again, at the right-hand edge of the picture.

The clutter of vehicles on the highway gives a good idea of the way in which accidents are produced. One motorcyclist has just pulled off to stop at the drinking-fountain by the roadside. His companion has pulled over and slowed down. This has caused the cars also to slow down, and the trucker has decided to take advantage of the situation and pass all three cars at once. Motorcycle, cars, and truck are thus, at the moment, all speeding abreast on a two-lane highway.

### 84 Highway and River

**US 40** An occasional airplane view demonstrates certain features of the highway better than they can be seen from the ground, although at the same time giving a fallacious impression, by displaying many features that a motorist cannot see. The picture is here taken, looking southwestward, from a plane flying about three thousand feet over the surface of the ground immediately below. The Truckee River, flowing from right to left, sweeps down in three great loops.

U. S. 40 keeping mostly to the left of the river, crosses it by a bridge in the middle distance, cuts across the loops, and shortens the distance greatly. The railroad, unable to climb grades as the highway does, follows the river closely.

The location is here a few miles east of the town of Truckee, California. In the foreground Prosser Creek flows into the Truckee through the sharply cut canyon. Martis Valley lies in the middle distance, with little Martis Creek flowing almost directly toward the observer. Beyond Martis Valley lies one of the Sierra ranges, and beyond it the large white expanse is Lake Tahoe under the cover of a low bank of fog. The main chain of the Sierra Nevada, largely snow-covered, lies in the distance. Most of this, of course, would be quite beyond the view of the motorist, and he would also be unable to see such features as the small lake on the left and Truckee airport at the right.

Actually, however, the motorist sees the landscape more intimately, and therefore sees much that the view from the airplane fails to reveal. For instance, at the left in the foreground, the road skirts closely beneath a picturesque flow of basalt not to be appreciated from the plane.

Historically, this scene is of interest as showing the spot at which the Stevens Party emerged from upper Truckee Canyon. To advance three miles here they must have been forced to cross the river at least three times, and they would have expended half a day at making this distance, or more, if the crossings had proved difficult.

The building with the turnout just beyond the bridge is one of which the motorist is likely to see entirely too much. This is the California State Quarantine Station, where all westbound motorists must stop and open their baggage for inspection, on suspicion that they may be carrying plants or vegetables which harbor insect pests.

## ⑧⑤ Donner Pass

🛡40 Looping upward, U. S. 40 ascends the precipitous eastern slope of the Sierra Nevada. The picture is taken from high up on the southern slope of Donner Pass, well above the railroad tracks. The time is a September afternoon with the westering sun casting heavy shadow-blotches. The sheen of the superb Sierran granites, however, still shows where the full light strikes. The day is hazy, with forest-fires in the distance, so that the crags of Castle Peak seem farther away than their actual five miles. The highest tip of these crags reaches an altitude of 9139. The highway crosses the summit of the pass, just outside the edge of the picture to the left, at 7135. These heights are small compared as with those in Colorado.

Yet the country around Donner Pass seems more alpine than that around Berthoud. This is a result partly of steeper slopes, partly of more recent glaciation which has left the granites exposed, and partly of a more snowy climate which prevents the growth of trees above 7000 feet except in favored spots. It can be noticed in the picture that the trees seem to be growing in the shadow. This is as much as to indicate that they are growing in spots where they are sheltered from the prevailingly western wind. Where trees manage to grow, they are most commonly red firs. The rest of the ground, where it is not bare granite, is generally covered with a thick low mat of alpine shrubs, looking from a distance like patches of moss.

Exactly where the emigrants made their crossing is now probably beyond sure determination. The author, working on the history of the Donner Party and also out of mere interest in tracing old trails, has scrambled around the pass on many occasions in the past twenty years, but the record has been so badly obscured by the later work of the builders of highways, railways, and pole-lines that he has been able to discover little. Always he has returned, not only having failed to find actual indication of the route, but also overwhelmed with amazement that wagons could ever have been taken across the pass anywhere. Presumably they were brought to the level spot to the right of the picture and were taken across to the level spot at the left of the picture, which would have served as a jumping-off place for the top of the pass. In between, it seems impossible that they could have been taken along the boulder-filled stream-course paralleling the present highway, or by the equally boulder-filled ravine spanned by the bridge, also impossible that they could have been hauled up the rough and steep granite slopes. Yet somewhere they went over.

About one mile of road shows in the picture. Although there happen to be no trucks passing at the moment, the heavy traffic is indicated by a bus and eight cars, in addition to the one parked at the observation post.

Just at the left, paralleling the curve of the highway, a short stretch of the old Dutch Flat and Donner Lake Road still shows. The bend of the road in the foreground is known as Windy Point to the snow-plow men. On account of the conformation of the ground in that vicinity, it is the meeting place of all the winds, and its clearance during the winter storms is a perpetual challenge.

**86** Snow Scene

(40) U. S. 40 is a great road for winter sports, particularly in the vicinity of Berthoud and Donner passes. The location is here just east of Soda Springs, California, a few miles down the western slope from Donner Summit, at about 6000 feet.

There is no cut in the roadway, and the banks consist entirely of snow. Nevertheless, conditions may be described as moderate, for this area. The snow-stakes are still standing up high, and have not had extra stakes attached to their tops.

The total annual precipitation is here about 46 inches, and most of this generally falls as snow. Since one inch of precipitation is roughly represented by ten inches of snow, the total fall for a year may add up to 30 feet. This fallen snow melts somewhat and always consolidates. Nevertheless, the depth even where there are no drifts, often has to be measured in yards rather than in inches. Along the road, there is of course an extra depth of snow, because it is piled up on either side where thrown out of the line of the road by the rotary snow-plows.

At the time of the taking of this picture there had just been a short but severe snow-flurry. Either falling or blowing snow is still obscuring the ridge in the background. The push-plows have already been along the road and have cleared the snow back from the center. It is still piled along the edges, and will later be thrown back from the road by the rotary plows. This spot is only about twenty miles from the Truckee River bridge at Verdi, where the picture taken on the same day shows only a slight skim of snow.

The trees are red firs, growing under severe alpine conditions. Their thoroughly plastered condition is evidence of very recent snow. With a little sun or wind the snow-loads will begin to slide off the naturally drooping branches, and the trees will be freed again. To the right, however, above the radiator ornament of the car, may be seen what is sometimes known as a "snow-ghost." This is a small tree that has become almost completely encrusted, is beginning to lean over, and may eventually arch down clear to the ground.

In this area U. S. 40 was blocked by heavy snows for four weeks in January-February, 1952.

## ⓼⃣ Forest Primeval

🛣️40 About a quarter mile east of its junction with California 20, U. S. 40 passes through a very small area that may be claimed as the only bit of big forest still left along this particular highway from coast to coast. These trees should by all means be preserved, but on the contrary, in all probability, will soon fall before the loggers. Why they have even escaped thus far is something of a mystery—perhaps because the area is actually very small, and because the large boulders make logging somewhat difficult. With the present price of lumber, however, such prime trees cannot long expect to escape the saw, unless especially preserved. Although within the boundary of the Tahoe National Forest, this particular area is not federal property.

The largest trees in the picture are Jeffrey pines. This variety of western yellow pine is distinguishable from the common ponderosa by the more longitudinally marked bark pattern. The tree at the left is approximately five feet in diameter and 150 feet tall. It contains—and this will be its downfall—about ten thousand board feet of lumber. Being probably over 300 years old, it may have been a seedling, or even a fair-sized young tree, on the day when the Dutch landed at Sand Hook and established the eastern end of U. S. 40.

This tree may actually be well over three centuries in age, for at this altitude (5500) growth is by no means as rapid as it is a thousand feet or two lower.

The smaller tree just to the right of the largest one is an incense cedar, displaying its identity by its wholly different bark pattern.

From the point of view of forest management these trees are actually ripe, or overripe, for cutting. They are adding only a very small yearly increment of growth, and for material economy it would be better to remove them and let the young trees start growing. The location of these trees along a highway, however, makes them inherently much more valuable for scenic purposes than for lumber.

The forest floor is strewn with large granite boulders, between which is a growth of bushes and small trees. As is customary in primeval forest, the thick shade thrown by the large trees prevents heavy underbrush, and the forest is fairly open, easily traversed on foot or even on horseback.

The excellent tree-growth and the good stand of bushes, in spite of the shade, are evidences that this area receives the heaviest rainfall of any on U. S. 40. Here, two-thirds of the way up the western slope of the mountains, the precipitation is about 55 inches annually.

About a mile from here the streamliner was snowed in during the heavy snowstorm of January, 1952.

# ⓼⓼ Emigrant Gap

🛡40 Eastward from Nyack Lodge, some high peaks of the Sierra Nevada, though not the main ridge, are in view. Old Man Mountain (7775) rises against the skyline slightly to the left of the center. It is a typical Sierran dome of gray granite. Red Mountain, to the right, stands out more darkly, being largely forested and showing little effect of glaciation.

The general line between glaciated and unglaciated country shows up strikingly in the picture. At the higher levels, particularly toward the left, large gray areas indicate the granite surfaces that have been left exposed by glaciation. Closer at hand the thick dark growth of forest shows land from which the soil has not been thus removed.

The picture serves as an illustration of the way in which even modern highways follow watersheds. The narrow ridge, of which highway and railroad both make use, separates the headwaters of Bear River on the left from those of the American River on the right.

Historically, as the name of Emigrant Gap would indicate, the spot is of interest. The Stevens party, followed down Yuba River, and then came across into Bear Valley through the wooded country to the left of the ridge in the center of the picture. In the summer of 1845 they brought their wagons along this route. Later in that year, however, a better route was worked out, and the emigrant trail thereafter came down through the meadow (Carpenter's Flat), showing faintly at the right of the picture. From that point it came on, and passed through the watershed ridge at a low spot which came to be known as Emigrant Gap. So much work has been done in the construction of the railroad and the highway that the true location of the gap must be considered doubtful. Possibly it was at the spot where the highway crosses under the railroad. More likely it was at a gap now occupied by a large railway-fill, just beyond the farthest point where the railroad appears in the picture.

After the emigrants had got their wagons to this crest, the custom was to unyoke the oxen and to let the wagons down the steep slope by means of ropes snubbed around trees. This was the last really difficult spot on the emigrant road, and once the wagons had reached Bear Valley, the oxen were usually given some time to rest, and there was a feeling of general relaxation.

The railroad, although now doubletracked, still follows the line of the original transcontinental route put through in the sixties. The highway also follows the line of the old Dutch Flat and Donner Lake Road. The highway, however, cries out for relocation. As the picture clearly indicates, traffic is heavy (more than 3000 vehicles daily) and the blind curve is much too sharp for safety. There is, however, a fairly good view westward from this spot, and the white convertible in the picture is taking advantage of this to get by the lumber truck.

This truck is carrying a load of Douglas fir logs, of fairly good size, but of rather poor quality as regards straightness. In view of the fact that the forests in this region have been hacked and burned continuously since the first putting through of the railroad, it is remarkable that they can still supply any timber even as good as this.

The half illegible inscription on the face of the rock above the highway read originally "Christ died for our sins." It is a remnant of the time when the old road was at a much higher level.

Coarse roadside weeds appear in the foreground. One of these is just shooting up into a flower stalk.

**89** The Diggings

(US 40) Just east of Gold Run the highway shoots out into an open straightaway, cut at the base of high red cliffs of gravel. These are the old gold diggings, so-called, although they were not dug in the ordinary sense but were washed out by great hydraulic hoses that worked here in the sixties and seventies.

These hydraulic workings, a far cry from the original Forty-niner "with his gold pan on his knee," transformed the whole face of the country, and piled so much debris upon the agricultural lands of the valleys that this method of mining was eventually restrained by law. At this particular point, however, the hoses ceased work because they had arrived at the right-of-way of the Southern Pacific Railroad, which here runs as across a causeway, at the top of the cliff—a similar one having been cut away also on the other side of the railroad. Until the 1930's the highway also ran on the top of the cliff, closely paralleling the tracks.

The long-continued sluicing removed all the top-soil and even the overlying gravels, clear to bedrock, over many square miles. Since that time, however, the country has had seventy years in which to recover, and the drama of that recovery is well displayed in the picture. At the top of the cliff, where the topsoil was never washed away, we have ordinary second- or third-growth Sierra forest, chiefly of ponderosa pine. Along the very lip of the cliff in the full sunshine grows an understory of manzanita bushes. But to the left of the road, in the devastated area, the ponderosa pines have also managed to take over. Although some bare spots still show, these are yearly becoming fewer and smaller, and in another thirty years the two sides of the highway may be superficially indistinguishable. One hardy little pine has even managed to establish a footing halfway down the face of the cliff.

When the highway was put through, the lower part of the cliff was somewhat cut back, thus producing the two curious "hanging valleys" emphasized by their growth of bushes, which run halfway down the face of the cliff and then are cut off. Since the building of the highway a certain amount of talus has weathered off the cliff, and just at the end of the straightaway one may notice that a few vigorous pines have already taken root upon this recent slope.

## ⑨⓿ Coast Range

🛡️40 In the fall of 1949 this section of U. S. 40, a few miles to the northeast of Vallejo was being transformed into a four-lane freeway, fully modern in all respects. The need for the transformation is clear enough from the picture.

The first truck, with trailer, is hauling a cargo of large and well-filled sacks toward San Francisco, probably rice from the Sacramento Valley. The second truck declares its business in no uncertain terms. One slow truck would be bad enough; one behind another constitutes an almost complete road-block. Fifteen cars are piled up behind the trucks, and the first driver is just deciding to take a chance at getting by. One rugged, or extreme, individualist has decided to dare the traffic-police by going over on the side of the highway not yet open for traffic. Let us hope that he is some local resident who knows what he is doing.

This is a good example of a dominating highway. The gigantic fill has changed the whole appearance of the country, the opposing hill has been cut away for rock and gravel. What was once apparently the bed of a small stream has been completely made over, and the stream has been diverted through a large culvert and into the drainage-ditch—emphasized by the shadow—on this side of the highway.

In the foreground and in the lower left corner runs the standardized fence, of redwood topped with barbed-wire, that lines the freeways of California, preventing access except at established intersections. The time of day—about five in the afternoon, as the stretched-out shadows would indicate—accounts for some of the clutter of cars, but this road always carries heavy traffic. Its daily average for 1950 was about 13,000, half as much as that carried by the six-lane highway pictured earlier. Yet until that year all this traffic was being handled on two lanes.

The muted tones of the picture provide a good rendering of the Coast Range in the autumn. The sky is cloudless and hazy blue. On the hills the grass is ripe (say Californians) or dead (say Easterners), and the color of the hills is golden (say Californians) or brown (say Easterners). In any case there is little contrast.

In the background and at the right the smooth, though sometimes steep, contours of the Coast Range are visible. This is excellent grazing-country. Typical of the steeper hillsides are the closely spaced parallel paths beaten down by cattle, faintly observable at the right of the picture.

The two clusters of native live-oaks are characteristic of the Coast Range. Just why these trees grew in such clumps, often—as here—in the steepest and rockiest places, is a question difficult to answer surely. According to one theory, the Indians burned off the country during the dry season, and the oaks were thus restricted to places where fire was less destructive. Another possibility is that in steep and rocky places the trees are not so subject to the depredations of grazing animals.

The car parked beyond the highway belongs to the photographer, who crossed the pavement, climbed the fence, and went a little way up the hill for his picture.

## ⑨¹ San Francisco Bay Bridge

(US 40) The world's greatest bridge, and the proud city to which it leads, together form a fitting climax to the westbound highway. The bridge is photographable from so many angles that there is an embarrassment of riches. From certain points upon its course the motorist can look out toward the Golden Gate, its bridge, and the ocean beyond. The view here used, however, emphasizes the highway itself, and the San Francisco skyline beyond.

The picture was taken from a moving convertible in an eastbound-lane on a Sunday morning, and shows the six-lane upper deck of the bridge in a quite uncharacteristic state of desertion.

The great suspension bridge itself is a striking example of the best in modern architecture, startling in the simplicity of its pattern. In the upper left-hand corner appear first the two wires serving as hand-guards for the workers. Below these wires is the great main cable. Dropping from this are the cables upon which the bridge is hung directly, presenting patterns of double pairs. Next in view on the bridge is the heavy rail supported upon the solid steel parapet. This may be considered one of the few blemishes in the architecturally magnificent whole, for to an unnecessary degree the parapet and railing produce an effect of heaviness and interfere with the view. Beside the parapet runs a narrow footwalk, serving only for workmen, since pedestrians are not allowed on the bridge. Closer, are the three westbound lanes, wholly deserted; then, the narrow painted dividing-strip and the eastbound lanes. To the right, one of the lampposts, supplying yellow sodium light, rises above the line of the parapet.

In addition to the six lanes on the upper level, the lower level has three lanes for trucks and buses, and also carries a double tracked railroad for interurban electric trains.

Eighty thousand motor vehicles cross this bridge daily. The yearly total thus approaches thirty million—more than cross any bridge in the country, even the toll-free Williamsburg Bridge across New York's East River. In addition, there is the heavy traffic of electric trains. This is probably the most heavily traveled bridge in the world.

No city along U. S. 40 equals hill-built San Francisco in the bold picturesqueness of its skyline. At the left is the downtown district, with the Pacific Gas and Electric Building closest of the skyscrapers. At its right, rise the Shell Oil and the Standard Oil buildings. Towering from the sharp eminence of Nob Hill, the pyramid-topped Mark Hopkins Hotel surpasses the skyscrapers of the business district, in its turn to yield to the apartment-house rising from a somewhat higher level. Separated by a dip, the high buildings to the farther right rise from the top of Russian Hill.

As the cloudswept sky and sharply etched skyline indicate, it is a winter day with a north wind blowing.

# 92 End 40

U. S. 40 ends in San Francisco, on Harrison Street at the corner of Tenth.

After the beauty of fields and forests, after the grandeur of mountain and desert, after the dignity and power of the city as seen from the Bay Bridge itself—here, after three thousand miles, in full anticlimax, is the end, in a scene that is devoid of beauty or grandeur or power. Even the sign announcing the end of the two great highways is slightly knocked askew.

Others, before, have noted the anomaly, and have commented upon the irony of a great highway that crosses the whole of the United States, to end at a junk yard. To be a little more charitable, and also more accurate, we may at least say that U. S. 40 ends, characteristically, at a service-station, or even two service-stations.

Or, we may also say, it ends from the highway-planners' point of view, where it meets U. S. 101, as the sign here shows. This in itself may be considered no bad ending, for U. S. 101, a Canada-to-Mexico route, is another great highway.

Yet even the scene here portrayed is not wholly without its interest. To begin with, we should state that the desertion of the street is not typical, but that the picture was taken on a Sunday morning, to allow the photographer to stand in the center of the street without endangering his life and also to provide a clear view of the signs. Two of these signs are characteristic of San Francisco—the newspaper pullover, and the tow-away zone. This tow-away law, highly unpopular with careless motorists, is a recent device by which the city is striving desperately to keep traffic moving.

As other characteristic touches of an American city we may note the carefully suspended wires of the trackless trolley, and the two large billboards, one of them with its carefully placed floodlight, both turning their backs toward the photographer.

By being carried only a few miles farther, over already paved streets, U. S. 40 could be made to end at some magnificent location overlooking the Golden Gate and the Pacific Ocean. The transcontinental tourist could thus be made to feel that he had been brought to a climactic ending. By being carried only a half mile to the north, the highway could end at the Civic Center, in a scene of considerable urban grandeur.

Yet perhaps the present end is the most fitting of all. U. S. 40 is no swank boulevard, no plush parkway. If there is romance about it, this is the romance of the modern world, busy with its own affairs. U. S. 40 is the thoroughfare of a hundred thousand trucks and buses, and of a million undistinguished coupés and sedans, convertibles and station-wagons—driven east, driven west—for business or for pleasure, twenty-four hours a day, in a seven-day week. Most of all, U. S. 40 is a working road, and when it dead-ends into U. S. 101, its job is finished. Why should we seek an artificial climax? Work done, why should a road—or a man—make heroic gestures? *End — U. S. 40.*

# A FEW REFLECTIONS

A transcontinental journey over U. S. 40—or over any other coast-to-coast road—should be, for any thinking person, a somewhat sobering experience. He has traversed more than one-eighth of the circumference of the globe, the whole breadth of a great nation and of one of the earth's two great land-masses.

In so doing, he should at times have turned his thoughts to the past, to the present, and to the future. The past and the present of the road I have already tried to demonstrate by means of text and pictures. Perhaps a few speculations as to the future would be in order.

U. S. 40—crossing the country by a central route, making use of natural passageways, touching many long-established cities, already representing a tremendous financial investment—seems destined to remain an important line of travel as long as our civilization remains. As long, moreover, as our civilization develops and amplifies, U. S. 40—by whatever name it may be called—will also develop and amplify. In the distant future it may be, indeed, superseded by some other means of travel, possibly travel by air. If the present, however, indicates the immediate future, we may say that its development is progressing along three lines. It is becoming a four-lane highway; it is being transformed into a freeway; it is by-passing cities and towns.

At present about one fifth of U. S. 40, including its sections in cities, is of more than two-lane width. Most of this is of four-lane, but there is a little of dangerous three-lane, and a little of six-lane or wider. Delaware and Indiana are far in the lead, followed by California and Ohio.

The freeway, or limited-access highway, seems to be one of the answers to allow high-speed traffic with safety. Otherwise, every new highway rapidly throttles itself, by providing the transportation that encourages so-called "ribbon development" to grow up along it. To prevent the growth of these shoestring-like towns, parasitic upon the highway, the freeway seems to be the best method. As yet, however, there is comparatively little of such road. California here takes the lead, and the freeway now extending nearly all the way between Sacramento and San Francisco, seems to represent close to the ultimate in motoring. It possesses the advantages of a parkway, and yet passes through farmlands and natural countryside that have not been artificially prettified.

The by-passing of towns and cities is perhaps the most pressing need along the route. Of the larger cities, only St. Louis has handled this problem with some degree of adequacy. Next come the East Bay cities of California. In contrast the highway plugs squarely through the centers of Baltimore, Columbus, Indianapolis, Kansas City, Denver, and Salt Lake City. Although a large number of the smaller towns have been by-passed —notably by the freeways in California and by the new four-lane construction in Missouri—the greater part of the work still remains to be done.

As I have already stated, no observant man can well complete the run on U. S. 40 without some sobering thoughts. Most sobering, perhaps, is the sense of power. There are indeed long stretches of desert and mountain and scrubby woods, but by and large the productivity of a nation that is at the same time almost a continent becomes gradually overpowering, as one looks at it along both sides of the highway. The people of the United States have been granted a natural heritage such as perhaps no other people have ever been granted, and they have exploited it materially. Mile after mile, hundred-mile after hundred-mile, stretch the farmlands, interspersed with mines and oil fields, dotted with towns and great cities full of manufacturing plants. A jingo imperialist would be justified in feeling drunk with power.

One may indeed ask, "Is not this heritage now at its full peak of production, ready to decline? Has not the soil been depleted of its riches? Are not the oil-fields and the mines now at their peak, ready to decline?" It may be. One passes gullied and eroded hillsides. There is a surprising amount of land that was once farmed, and is now going back into forest, depleted. There are abandoned mines, and oil-fields where the pumps are no longer going.

Yet on the whole the argument would seem to work in the other direction. The wastage itself, though appalling, is indicative of a kind of greatness. Only a supremely prosperous people could afford to waste so much—to let land revert to unproductiveness, to be careless of erosion, not even to practise forestry.

Take, as an example, one small item. On both sides of the pavement along most of the course of U. S. 40 grass grows profusely. In almost any other country there would be some provision by which this grass could be fed to animals. Either they would be tethered there and allowed to graze, or the grass would be cut. Along U. S. 40 these countless potential bales of hay are nearly all allowed to go to waste.

If—say about the year 1937—one Adolf Hitler could have been spirited away and taken upon a tour across U. S. 40, what might have been the effect upon the history of the world? Would he ever have let himself become embroiled in a war into which the United States was in the long run almost certainly to be drawn? Did Hitler, who had never traveled much beyond the confines of central Europe, have any real idea of the power of the United States? Such a conception can hardly come from reading figures in books. It comes, in an entirely different way, when one drives at the speed of the modern automobile, day after day, through highly populous and amazingly productive country.

No matter what the final judgment should be, a journey over U. S. 40 is at least highly instructive. There is an advantage in seeing a cross section. There is an advantage even in the present imperfect state of the highway that enables one to plow through the centers of so many of our cities. By following U. S. 40 you do not travel a parkway, and do not wander from national park to national park, seeking the spectacular in scenery. Instead, accepting the commonplace along with the spectacular, seeing the people and the country too, taking the good with the bad and the beautiful with the ugly, you gain some balanced impression of the United States of America.

**THE NAMES
AUTHOR'S NOTE**

# THE NAMES

On any important highway hundreds of signs have been erected to let the motorist know what towns he is approaching and what streams he is crossing. Some of the more important summits are also marked. Of the states along U. S. 40 New Jersey has been the most meticulous, but the posting is everywhere adequate.

These names, constantly forced upon the attention, arouse the motorist's curiosity, and lead him into constant speculation. To anyone who has a little background in the process of naming, the signboards can be a source of continual interest. The history of any region can be read in terms of its nomenclature.

Names are often classified according to the motivations that lead to their application. The basic motive is, of course, always to identify the particular place by distinguishing it from others. To accomplish this, one of the most common and primitive methods is simply description. Places along U. S. 40 thus bear such simple labels as Knolls, River Bend, Gumbo, High Hill, Big Springs, and Piney Grove. Not quite so obviously, The Rabbit Ears describes the shape of the rock formation; Steamboat Springs is derived from the chugging noise produced by an outflow of boiling water; from First View in eastern Colorado the Rocky Mountains are first seen; Saltair is close to Great Salt Lake. Description need not be merely of natural features, but may refer to a man-made situation. Thus arise such names as Junction City, Bridgeport, and Weskan, for the westernmost town in Kansas.

What may be called "relative description" occurs when a place is named by reference to some other place, as with Nine Mile Creek. A "distance-name" usually indicates that the stream-crossing was a certain number of miles from some place at which people usually set out, such as a stage-station or fort. By a relocation of the road, altering the distance, the

origin of the name may be obscured, and the name become a misnomer.

Only a step removed from simple description are the names that spring from association with people. May's Landing was, in a sense, a descriptive name as long as George May, the dock-builder, lived there. Of similar nature are Pop's Place, and Dad Lee's. The name may spring not from residence or ownership, but merely from some vaguer or more casual association. Thus Berthoud never owned the pass that bears his name and never lived there, but he discovered it. Indian Springs may indicate that some Indian or Indians lived there, or the name may merely record the finding of arrowheads or some other relics. Many Indian names of streams do not describe the stream itself, but merely show that a certain tribe once inhabited that region, as with Susquehanna, Potomac, Missouri, and Kansas.

Now and then, what seems to be a very curious name turns out to be merely something personal. An example is Pancake, in Pennsylvania, which suggests an interesting story until it turns out to be only the family-name of an early settler.

Another primitive type of name is that springing from an incident. Such a one is often hard to be certain about, for in the course of time the incident is forgotten, and people begin to think that the name is descriptive. Troublesome Creek, for instance, is almost certainly a name that records some particular trouble suffered there by an early traveler. Occasionally an actual record of the incident is preserved. Thus with Battle Mountain we actually have record of a fight with the Indians. A large number of animal-names are of this type. Elk River probably does not indicate that elk were especially numerous at that place, but more likely indicates that some early hunter happened to kill an elk there. Skull Creek undoubtedly records the finding of a skull. An interesting and apparently authentic case of incident-naming occurs at Kingdom City, Missouri. During the Civil War this district maintained neutrality in the local bushwacking, and so was humorously said to be a kingdom of its own.

Another class of names—extremely common for towns all the way across the United States—may be termed commemorative. These are named in honor of some important person, or reproduce nostalgically the name of some town "back east," or "in the old country," or grandiloquently echo some important name from history or from the contemporary world. In the eastern part of the United States the persons commemorated are often members of the English nobility, such as Lord Delaware and Lord Baltimore. Fort Cumberland, from which the city takes its name, was so called in honor of the Duke of Cumberland, son of George II, commander-in-chief of the British army. Farther west the names of American heroes begin to appear, and we have Washington, Claysville, West Jefferson, Lafayette, Marshall, Pocahontas, Tecumseh, Kit Carson, and Roosevelt. Knightstown was named, quite aptly, for one of the engineers who laid out the National Road. Winnemucca commemorates a local Paiute chief. A special class is formed by the saints' names, and one signpost along U. S. 40 in Missouri is probably unique in the United States as bearing simply the names of three saints.

Although many people have made fun of this commemorative naming, it actually records—often in nostalgic and sentimental terms—much of the history of settlement. Teutopolis and Vevay Park, for instance, indicate centers of German and Swiss immigration. Victoria, in Kansas, though it is now a German-Russian settlement, shows in its name that it was started as an English colony and christened for the reigning queen.

Commemorative naming from places follows similar lines. In 1806, because of the disturbances of the Napoleonic War, some people from the English island of Guernsey, off the coast of France, migrated to Ohio, and their old home is commemorated by Guernsey County. Its county-seat, however, commemorates the English Cambridge only at second-hand, having been immediately named because some of its settlers came from Cambridge, Maryland.

This repetition of names over the United States produces interesting, if somewhat confusing, effects. Thus U. S. 40 may be truthfully said to pass not only through Baltimore and St. Louis, but also through Washington, Philadelphia, Cleveland, Albany, and Detroit. In addition it passes through Hebron and Troy, and traverses such interesting foreign places as Malaga, Odessa, Toulon, Naples, Amsterdam, Strasburg, Glasgow, and South Vienna, and even whole countries, such as Bavaria and Brazil. Also—without ever doubling back or entering New York—U. S. 40 passes twice through Manhattan.

Many commemorative names slide over into the next class, which may be called euphemistic. These names attempt to give the new town a good start by bestowing an attractive name upon it. Golconda—now a highway station, but once a mine—reproduces the name of a district in India associated with rich diamond mines. In Nevada the riches would have taken the form of gold and silver, and the name may have been given to help the sale of stock or merely to bring good luck. Silver Zone Pass is another euphemistic miner's name. As agricultural counterparts, we have Richland, Grainfield, and Fruitland. Vaguer, but still suggestive of the happy side, are Pleasantville, and Fairview.

Many names are secondary in their derivation, that is, they are derived from other names. An obvious example is the Smoky Hill River, which the Kansas road-signs spell Smokey. The derivation Smoky Hill itself is disputed. It may well be an incident-name, recording a grass fire seen from a distance.

The process known as folk-etymology is especially productive of interesting names. Wheeling looks like an English name, but it is probably based upon an Algonquian Indian word meaning *place of the skull*. Weimar, in California, comes from an Indian chief whose name was spelled Weima, Weimah, or Wimmer; it was later confused with the name of the German City, and was so spelled.

The preservation of foreign words in names often makes them seem more mysterious and unusual than they really are. Although U. S. 40 runs for a short distance through territory settled by the Swedes and the Dutch, their languages have been replaced, and probably no signpost along the highway now bears a name directly derived from those early settlers. There are, however, many Indian names. As far as Missouri, the highway runs through territory which was inhabited by Algonquian-speaking tribes. The meaning of many of these names is uncertain, but when they can be translated, they generally turn out to be simple descriptive or incident names. Thus Conococheague means something in the general nature of "far away," and was probably applied because the creek lay at a considerable distance from some particular Indian village. The word-element *con* certainly means "long," and is preserved in other names such as Connecticut, and Conemaugh. An occasional Iroquoian name intrudes into this area. Susquehanna originates from an Iroquoian tribe, and Ohio means roughly, "fine river."

Farther west, other Indian languages supply the native names. Most important along U. S. 40 is the territory of the tribes speaking the related Ute-Shoshone-Paiute dialects. From the Ute comes, for instance, Timpanogos, which means approximately "rocky river," and was applied secondarily to the mountain.

In the Middle West many names are still preserved from the original French occupation. Such a one is Terre Haute, meaning merely "high land" indicating a place along the river that was safe from floods. Loutre River preserves the French word for otter. Vermilion River was not so named from the color of the water, but records the existence of a red mineral, used by the Indians for painting, which the French thus termed. Embarrass River is not an incident name, but is derived from the literal meaning of a word in French and indicates an "em-barred," stream—that is, one that was blocked by fallen and floating trees. Auxvasse River combines a French article-preposition with a noun, and means "at the salt lick." The grammar—as not infrequently happens with names—has gone to pieces, for the gender, number, and spelling are all questionable. Sni-a-bar Creek shows a French name become almost undecipherable. The first syllable is derived from a dialect word *chenail*, meaning "channel." The name was originally in some such form as *chenail-à-barre*, and the meaning was thus about the same as that of Embarrass River.

Spanish names occur along U. S. 40 in California. Rodeo indicates a place where cattle were rounded up. Pinole, originally an Aztec word, is probably an incident name, based upon the giving of some *pinole*, a sort of meal, to some Spanish soldiers by the Indians.

Many foreign names have of course been given in more recent times, and do not indicate an early settlement. Thus Boca, in California, is good Spanish, meaning *mouth*, and is aptly applied, because at this point there is the mouth of a stream. It does not date back to any Spanish explorer, however, but was applied about 1867, during the building of the railroad.

The names across the country also indicate different dates and different levels of civilization. Generally speaking, streams were named before towns, and when the two bear the same name that assumption can generally be made. There are exceptions, particularly in the case of very small streams, which often were not named at all in early times. But in general, as with Sacramento and Truckee, the town is the later. When a state and a river bear the same name, the state was named from the river, as with

Ohio, Illinois, Missouri, and Kansas. Colorado is an unusual case, but not really an exception.

Names incorporating the words for animals, such as bear and elk, also indicate early naming. On the contrary such a name as Phoneton, in Ohio, suggests very recent origin, and was actually so-called because of being a test-station and transmitting-center for telephone lines.

Sometimes a once good name becomes obnoxious, or merely humorous, and a movement arises to change it. Occasionally the movement is successful, especially with recently founded towns. More often, the citizens stubbornly stick to the original name. A curious halfway change is displayed on a sign in Indiana, which records the name of the village as East Germantown, but that of its post-office as Pershing. There was obviously a clash of old and new loyalties and sentiments at the time of World War I.

When the saying, "Where's Elmer?" became common, and humorous, the town in New Jersey did not change, but with good spirit erected a sign, "Here's Elmer."

By and large, an alert motorist can get a good deal out of watching signposts. Most guide-books give some information about names, and inquiry at a service-station often yields interesting results, which should be taken with a grain of salt. Roadmarkers erected by the state of Pennsylvania give the derivation of most of the town-names within its boundaries. But even without authoritative sources of information, mere speculation itself can often be interesting.

# AUTHOR'S NOTE

The book would have been impossible to prepare without the co-operation of many organizations and individuals. To them all, some of whom may at this date have even been lost from my records and have lapsed in my memory, I return my thanks.

I must mention: the Bureau of Public Roads, the American Association of State Highway Officials, and the highway commissions or departments of all the states through which U. S. 40 passes. From them I have received much information, historical and statistical. I have also availed myself of the co-operation of Pacific Intermountain Express, the Divisions of Architecture and Engineering and of Parks and Memorials of the State of Illinois, the St. Louis Chamber of Commerce, the Kansas City Chamber of Commerce, the Tahoe National Forest, the Esso Touring Service.

Various historical societies have furnished me with material, viz., those of New Jersey, Maryland, Western Pennsylvania, Missouri, Colorado, Frederick County (Md.), and Salem County (N. J.). Of libraries I must include the Illinois State Historical Library, the University of California Library, and the Bancroft Library.

A large number of individuals helped with the book by granting me admission to their premises for the taking of pictures and by supplying me with information on the spot. In most cases I did not learn their names, and I can thank them only *en masse*. Courtesy and friendliness are striking characteristics of citizens generally along this cross section of the United States. As a note of grim humor, I should add that I was twice suspected of being a Communist spy.

Of particular persons I am especially indebted to my wife, who made one trip with me, and to my son, John H. Stewart, who was my companion on the other trip. My colleagues in the University of California have, as often before, given me essential aid. I must mention in particular: Annetta

Carter, Robert A. Cockrell, Harmer E. Davis, A. Starker Leopold, Ralph A. Moyer, and Winfield S. Wellington. Especially helpful, as correspondents, were Carl J. Rech of Cambridge, Ohio, and Annie C. Newell, of Salem, New Jersey. For various kinds of essential assistance I wish also to thank E. W. James, Lowell Sumner, Parker Trask, Jim Morley, and Jane Burton.

Less extensive than my debt to people is my debt to previous writings. Some of these are mentioned in the text. In general, there is a wealth of printed material about that part of U.S. 40 which has sprung from the National Road; very little, about the rest of it. Philip D. Jordan is the latest historian of the National Road. I have also used his predecessors Robert Bruce, A. B. Hulbert, and T. B. Searight, and the special studies of J. S. Young, Lee Burns, and C. L. Martzolff. For Maryland, Volume III of the *Maryland Geological Survey*, particularly Part III, by St. George L. Sioussat, is excellent; for California, the Centennial Edition of *California Highways and Public Works* (Vol. 29, Nos. 9, 10), particularly Chapter XI, by Stewart Mitchell. In other states the history of the route has had to be dug out of special articles, and original sources.

Among these should be mentioned the studies of G. C. Broadhead (Missouri), and of LeRoy Hafen and Margaret Long (Colorado). Many other studies are sufficiently identified in the text.

As regards the descriptions of particular pictures, I have used too many and too scattered sources to make their listing practicable. I should mention, however, the W.P.A. guide-books for all the states traversed.

As a good general introduction to the whole subject I have used *Highways in our National Life* (eds., Jean Labatut, and Wheaton J. Lane). Also useful is *Highway Practice in the United States of America*, issued by the Public Roads Administration. I must certainly acknowledge my great debt to N. M. Fenneman's two volumes on the physiography of the United States, the standard works on the subject.